Shrinkage Regression Estimators and Their Feasibilities

Masayuki Jimichi

関西学院大学研究叢書第 182 編

Kwansei Gakuin University Press

Shrinkage Regression Estimators
and Their Feasibilities

Copyright © 2016 by Masayuki Jimichi

All rights reserved.

No part of this book may be reproduced in any form or by
any means without permission in writing from the author.

Kwansei Gakuin University Press
1-1-155 Uegahara, Nishinomiya, Hyogo, 662-0891, Japan
ISBN 978-4-86283-228-3

To my family

Preface

Around the middle of the 20th century, the following interesting studies in statistics independently started:

- Stein (1956) showed that the sample mean vector for estimating the mean of the multivariate normal distribution is *inadmissible* if the dimension p is greater than or equal to 3, and that some estimator has a smaller *risk* than the usual estimator. This fact is called *Stein's phenomenon* or *paradox*.

- Kendall (1957) considered an orthogonalization of explanatory variables in a usual regression analysis and applied *principal component analysis* to regression analysis. He also conjectured that the method throws some new light on certain old but unsolved problems, namely, *variable selection* and *multicollinearities*.

- Hoerl (1959) treated a problem in *response surface design* and considered a restricted estimation. The problem was caused by multicollinearity and the method of analysis was called *ridge analysis*. The origin of the term *ridge* is the shape of the locus of the solution set for equations with a sphere restriction.

After these studies, Hoerl and Kennard (1970) proposed an estimator obtained by adding a small positive quantity (*ridge coefficient*) to the diagonal elements of the information matrix in the *normal equations*, and showed that the estimator has smaller error than the *ordinary least squares* (OLS) estimator with respect to the *total mean squared error* (TMSE). Their proposed estimator is called the *ordinary ridge regression* (ORR) estimator. They also gave a generalized version of the ORR estimator called the *generalized ridge regression* (GRR) estimator. The *principal component regression* (PCR) estimator was proposed by Kendall (1957). Marquart (1970) regarded it as a *generalized inverse* (g-inverse) estimator, and considered several properties of the PCR and ORR estimators in detail. Furthermore, he mentioned an estimator obtained by combining the PCR estimator and the ORR estimator. Later, Baye and Parker (1984) referred to this as the *r-k class* estimator and investigated it.

All the above-mentioned estimators have in common that they shrink the usual estimator. Thus, they are called *shrinkage estimators*, or *shrinkage regression estimators* when they are applied in regression analysis, and have been studied for about six decades. Gruber (1998) performed a wide-ranging historical survey of the shrinkage estimators with respect to the Stein estimator and the ridge regression estimator.

It has been mentioned in the above papers that if there exists multicollinearity among the vectors of the explanatory variables in a regression model, then several problems occur. These problems relate to the accuracy of the OLS estimator for estimating the regression coefficients. In this book, we will improve the estimating accuracy by using the shrinkage regression estimators, and also consider their feasibilities. Additionally, the shrinkage regression estimators have customarily been treated under a standardized model; however, it is important in applications that we consider the shrinkage regression estimators under the original (usual) model. There has been little previous work on these topics, and it is also difficult to analytically obtain the total mean squared error of the feasible version of the shrinkage regression estimators.

This book is organized as follows. Chapter 1 defines the usual linear regression model, and considers its standardized model and canonical form. The OLS estimators for each model are also obtained, and the relationships among regression coefficients in these models are considered. *Invariance* of the OLS estimator is also mentioned. Furthermore, multicollinearity is defined, and its harmful influences are considered.

Chapter 2 shows several shrinkage regression estimators for the standardized model, and *shrinkage-type regression estimators* for the usual regression model, respectively. Chapter 3 proposes several *mean squared error* (MSE) criteria for them, and compares their *total mean squared errors* (TMSEs). Some methods for choosing the number of principal components and the ridge coefficients are also discussed.

Chapter 4 gives a variety of algorithms for choosing the number of principal components and the ridge coefficients. *Feasible* versions of the shrinkage regression estimators are also considered.

Chapter 5 precisely calculates the exact moments and the MSE criteria of the feasible GRR estimator. Note that the results can be used as a "benchmark" for its improvement.

Chapter 6 verifies the MSE criteria of the shrinkage regression (-type) estimators by a *parametric bootstrap method*. We determine how to improve them by comparing the results of several Monte Carlo simulations under multicollinearity.

Finally, Chapter 7 considers the *linear basis function model*, which is used in the fields found in statistical learning and machine learning, and estimates the coefficients in the model by using the feasible GRR estimators in the *overfitting* problem. A numerical example based on a polynomial regression model is also given, and we obtain an estimate of the advantage of the feasible GRR estimators over the OLS estimator.

This book [1] is based on Jimichi's doctoral dissertation (2005), with some recent results added to it.

Masayuki Jimichi

Kobe, Japan
August 2016

---

[1] This book has been typeset by using LaTeX and the figures have been drawn by using Maple 2015, R 3.3.1, and WinTPic 4.1.0.

Acknowledgments

I would like to thank Professor Emeritus Shingo Shirahata of Osaka University for warm and friendly guidance. I would like to thank Professor Emeritus Nobuo Inagaki of Osaka University. He always has inspired me to study statistics. I would like to thank Professor Yutaka Kano of Osaka University for instructive guidance. I also acknowledge Dr. Yuji Sakamoto of Kobe University for many useful conversations over the last several years. Dr. Etsuo Hamada of Osaka University and Dr. Shuichi Nagata of Kwansei Gakuin University read the draft version of this book and gave me many valuable comments. I am very grateful to them.

We are happy to acknowledge the support of Mr. Naoya Tanaka of Kwansei Gakuin University Press.

The publication of this book was supported by a Research Grant for KGU Publication Subsidy from the Organization for Research Development and Outreach of Kwansei Gakuin University.

Contents

Preface	iii
Acknowledgments	v
Contents	vii
List of Symbols	xi

1 Notation and Models — 1
- 1.1 Usual Model … 1
- 1.2 Standardized Model … 2
- 1.3 Canonical Form … 2
- 1.4 Least Squares Estimators … 3
- 1.5 Invariance of Ordinary Least Squares Estimator … 4
- 1.6 Multicollinearity … 5
- 1.7 Alternative Regression Estimators … 7

2 Shrinkage Regression Estimators — 9
- 2.1 Ordinary Ridge Regression Estimator … 9
- 2.2 Generalized Ridge Regression Estimator … 9
- 2.3 Principal Component Regression Estimator … 10
- 2.4 $r\text{-}k$ Class Estimator … 11
- 2.5 Invariance of Shrinkage Regression Estimators … 11

3 Mean Squared Error Criteria — 15
- 3.1 Definitions of Mean Squared Error Criteria … 15
- 3.2 Mean Squared Error Criteria of Regression Estimators … 16
  - 3.2.1 OLS Estimator … 16
  - 3.2.2 ORR Estimator … 18
  - 3.2.3 GRR Estimator … 23
  - 3.2.4 PCR Estimator … 28
  - 3.2.5 $r\text{-}k$ Class Estimator … 31
- 3.3 TMSE Comparisons among Regression Estimators … 35
- 3.4 Basic Methods for Choosing Number of Principal Components and Ridge Coefficients … 39
  - 3.4.1 Number of Principal Component … 39
  - 3.4.2 Ridge Coefficients … 39

4 Feasible Shrinkage Regression Estimators — 43
- 4.1 Feasible ORR Estimator — 43
- 4.2 Feasible GRR Estimator — 44
- 4.3 Feasible PCR Estimator — 45
- 4.4 Feasible r-k Class Estimator — 45

5 Exact Moments of Feasible GRR Estimator — 47
- 5.1 Boundedness of Moments — 47
- 5.2 First and Second Moments — 49
- 5.3 Cross Moments — 56
- 5.4 Mean Squared Error Criteria — 60
- 5.5 Numerical Evaluations — 63
  - 5.5.1 First and Second Moments — 63
  - 5.5.2 Cross Moment — 63
  - 5.5.3 Monte Carlo Simulations — 69
  - 5.5.4 Relative Efficiency — 70

6 Estimation of MSE Criteria by Bootstrap Method — 73
- 6.1 Introduction — 73
- 6.2 Parametric Bootstrap Method in Regression Analysis — 74
- 6.3 Parametric Bootstrap Method for OLS Estimator — 75
- 6.4 Parametric Bootstrap Method for Feasible Shrinkage Regression Estimators — 76
- 6.5 Monte Carlo Simulations — 78

7 Applied Feasible GRR Estimation to Linear Basis Function Models — 85
- 7.1 Introduction — 85
- 7.2 Models and Estimators — 86
- 7.3 Exact Moments and Mean Square Criteria of FGRR Estimator — 90
- 7.4 Numerical Example: Polynomial Regression Model — 91

8 Concluding Remarks — 97

A Special Functions — 99
- A.1 Gamma Function — 99
- A.2 Beta Function — 100

B Lemmas for Moments — 101

C Data Sets — 107
- C.1 Data Generating Model and Data Sets — 107
- C.2 (Data Set 1) — 108
  - C.2.1 Usual Model — 109
  - C.2.2 Standardized Model — 110
  - C.2.3 Canonical Form — 111

C.3	(Data Set 2)	113
	C.3.1	Usual Model	113
	C.3.2	Standardized Model	114
	C.3.3	Canonical Form	116

Bibliography 119

Index 124

List of Symbols

Symbol	Definition
$\mathbb{N}$	the set of natural numbers
$\mathbb{R}$, $\mathbb{R}^n$	the set of real numbers, n-dimensional Euclidean space
$\mathbf{0}$, $\mathbf{O}$	zero vector, zero matrix
$\mathbf{Z}$, $\mathbf{X}$, $\mathbf{A}$	matrices of explanatory variables
$\mathbf{Y}$, $\mathbf{y}$	vector of response variables, vector of observed (realized) response variables
$\boldsymbol{\theta}$, $\boldsymbol{\beta}$, $\boldsymbol{\alpha}$	vectors of regression coefficients
$\boldsymbol{\varepsilon}$	vector of errors
σ^2	variance of errors
$\mathrm{GM}(\mathbf{Y}, \mathbf{Z}\boldsymbol{\theta}, \sigma^2 \mathbf{I}_n)$	Gauss-Markov setup model
$\mathrm{GMN}(\mathbf{Y}, \mathbf{Z}\boldsymbol{\theta}, \sigma^2 \mathbf{I}_n)$	normal linear models (Gauss-Markov models with normality)
$\mathbf{A}'$, $\mathbf{a}'$	transpose of matrix or vector
$\mathbf{A}^{-1}$	inverse matrix
$trace\, \mathbf{A}$	trace of matrix
$\mathrm{diag}(a_1, \ldots, a_n)$	diagonal matrix
$\mathrm{rank}(\mathbf{A})$	rank of matrix
$|x|$, $\mathrm{abs}(x)$	absolute value
$\|\mathbf{a}\|$	Euclidean norm of vector
$\mathrm{E}(\cdot)$, $\mathrm{E}_F(\cdot)$	expectation (under the distribution F)
$\mathrm{V}(\cdot)$, $\mathrm{V}_F(\cdot)$	variance (under the distribution F)
$\mathrm{bias}(\cdot)$, $\mathrm{bias}_F(\cdot)$	bias (under the distribution F)
$\mathrm{MSE}(\cdot)$, $\mathrm{MSE}_F(\cdot)$	mean squared errors (under the distribution F)
$\mathrm{TMSE}(\cdot)$, $\mathrm{TMSE}_F(\cdot)$	total mean squared errors matrix (under the distribution F)
$\Gamma(x)$, $B(x,y)$	gamma function, beta function

Chapter 1
Notation and Models

In this chapter, the usual linear regression model is given, and its standardized model and canonical form are also treated. The ordinary least squares (OLS) estimators for each model are derived, and the relationship among regression coefficients in these models is considered. Invariances of the OLS estimator are mentioned. Furthermore, the definition of multicollinearity is given, and harmful influences of it on estimating the regression coefficients are considered. Note that some alternative estimating methods for handling the problem of variance divergence caused by multicollinearity are discussed.

1.1 Usual Model

Let us consider the following *usual linear regression model*:

$$(\text{M}_Z) \qquad \boldsymbol{Y} = \boldsymbol{Z}\boldsymbol{\theta} + \boldsymbol{\varepsilon},$$

where

(N1) $\boldsymbol{Y}$ is an n-dimensional vector of *response variables*.

(N2) $\boldsymbol{Z} := [\boldsymbol{1}, \boldsymbol{Z}_p]$ is an $n \times (p+1)$ matrix of *explanatory variables* with $\text{rank}(\boldsymbol{Z}) = p + 1 (< n)$, $\boldsymbol{Z}_p := [\boldsymbol{z}_1, \ldots, \boldsymbol{z}_p]$, and $\boldsymbol{1} := [1, \ldots, 1]' \in \mathbb{R}^n$.

(N3) $\boldsymbol{\theta} := [\theta_0, \boldsymbol{\theta}_p']'$ is a $(p+1)$-dimensional unknown vector of *regression coefficients* and θ_0 is the *intercept*.

(N4) $\boldsymbol{\varepsilon} := [\varepsilon_1, \cdots, \varepsilon_n]'$ is an n-dimensional vector of *errors*.

Assume that the errors ε_i $(i = 1, \cdots, n)$ are *independent and identically distributed* (i.i.d.) random variables with common distribution F, and their mean and variance are

$$\text{E}_F(\varepsilon_i) := \int_{-\infty}^{\infty} \epsilon_i dF(\epsilon_i) = 0,$$

$$\text{V}_F(\varepsilon_i) := \text{E}_F(\varepsilon_i - \text{E}_F(\varepsilon_i))^2 = \text{E}_F(\varepsilon_i^2) = \int_{-\infty}^{\infty} \epsilon_i^2 dF(\epsilon_i) =: \sigma^2 < \infty,$$

where σ^2 is an unknown parameter, and $\text{E}_F(\cdot)$, $\text{V}_F(\cdot)$ are the expectation and variance under the distribution F, respectively. Hence, the mean vector and the variance-covariance matrix of $\boldsymbol{\varepsilon}$ are

$$\text{E}_F(\boldsymbol{\varepsilon}) = \boldsymbol{0}, \quad \text{V}_F(\boldsymbol{\varepsilon}) = \sigma^2 \boldsymbol{I}_n, \qquad (1.1)$$

where $\mathbf{0}$ is the n-dimensional zero vector, and $\mathbf{I}_n := \mathrm{diag}(1,\ldots,1)$ is the identity matrix of order n. Therefore, the mean vector and the variance-covariance matrix of $\mathbf{Y}$ are

$$\mathrm{E}_F(\mathbf{Y}) = \mathbf{Z}\boldsymbol{\theta} = \theta_0 \mathbf{1} + \theta_1 \mathbf{z}_p + \cdots + \theta_p \mathbf{z}_p, \tag{1.2}$$
$$\mathrm{V}_F(\mathbf{Y}) = \sigma^2 \mathbf{I}_n. \tag{1.3}$$

The set $(\mathbf{Y}, \mathbf{Z}\boldsymbol{\theta}, \sigma^2 \mathbf{I}_n)$ is called the *Gauss-Markov setup* (e.g. Rao (1973)), and then the model (M_Z) is denoted $\mathrm{GM}(\mathbf{Y}, \mathbf{Z}\boldsymbol{\theta}, \sigma^2 \mathbf{I}_n)$.

1.2 Standardized Model

The following model is called the *standardized model*:

(M_X) $\qquad\qquad\qquad \mathbf{Y} = \mathbf{X}\boldsymbol{\beta} + \boldsymbol{\varepsilon},$

where $\mathbf{X} = [\mathbf{1}, \mathbf{X}_p]$, $\boldsymbol{\beta} = [\beta_0, \boldsymbol{\beta}_p']'$, and $\mathbf{X}_p = [\mathbf{x}_1, \ldots, \mathbf{x}_p]$ is the standardization of $\mathbf{Z}_p$. That is,

$$\mathbf{X} := \mathbf{Z}\mathbf{T}^{-1}, \tag{1.4}$$
$$\boldsymbol{\beta} := \mathbf{T}\boldsymbol{\theta}, \tag{1.5}$$

where $\mathbf{T} = \begin{bmatrix} 1 & \mathbf{m}_p' \\ \mathbf{0} & \mathbf{S}_p \end{bmatrix}$, and thus $\mathbf{T}^{-1} = \begin{bmatrix} 1 & -\mathbf{m}_p' \mathbf{S}_p^{-1} \\ \mathbf{0} & \mathbf{S}_p^{-1} \end{bmatrix}$ with

$$\mathbf{m}_p := [m_1, \ldots, m_p]', \quad m_j := \frac{1}{n}\sum_{i=1}^n z_{ij}, \ j = 1, \ldots, p,$$

$$\mathbf{S}_p := \mathrm{diag}(s_1, \ldots, s_p) \text{ and } s_j^2 := \sum_{i=1}^n (z_{ij} - m_j)^2, \ j = 1, \ldots, p.$$

Note that the *information matrix*

$$\mathbf{X}_p' \mathbf{X}_p = \begin{bmatrix} 1 & \cdots & r_{1p} \\ \vdots & \ddots & \vdots \\ r_{p1} & \cdots & 1 \end{bmatrix}$$

is in the correlation form, where r_{ij} is the correlation coefficient between the ith and jth columns of $\mathbf{Z}_p$ in the model (M_Z). Note also that $\mathbf{1}'\mathbf{X}_p = \mathbf{0}$.

1.3 Canonical Form

The following model is called the *canonical form*:

(M_A) $\qquad\qquad\qquad \mathbf{Y} = \mathbf{A}\boldsymbol{\alpha} + \boldsymbol{\varepsilon},$

where
$$\mathbf{A} := \mathbf{X}\boldsymbol{\Gamma}, \tag{1.6}$$
$$\boldsymbol{\alpha} := \boldsymbol{\Gamma}'\boldsymbol{\beta}, \tag{1.7}$$

and $\boldsymbol{\Gamma}$ is an orthogonal matrix of order $(p+1)$ such that
$$\boldsymbol{\Gamma}'\mathbf{X}'\mathbf{X}\boldsymbol{\Gamma} = \boldsymbol{\Lambda} = \mathrm{diag}(n, \lambda_1, \ldots, \lambda_p),$$
and
$$\mathbf{X}'\mathbf{X} = \begin{bmatrix} n & \mathbf{0} \\ \mathbf{0} & \mathbf{X}_p'\mathbf{X}_p \end{bmatrix}, \quad \boldsymbol{\Gamma} = \begin{bmatrix} 1 & \mathbf{0} \\ \mathbf{0} & \boldsymbol{\Gamma}_p \end{bmatrix}.$$

Note that $\{\lambda_1, \ldots, \lambda_p\}$ are the eigenvalues of the information matrix $\mathbf{X}_p'\mathbf{X}_p$ with
$$\lambda_1 \geq \cdots \geq \lambda_p (> 0). \tag{1.8}$$

The matrix $\boldsymbol{\Gamma}_p$ is an orthogonal matrix of order p such that
$$\boldsymbol{\Gamma}_p'\mathbf{X}_p'\mathbf{X}_p\boldsymbol{\Gamma}_p = \boldsymbol{\Lambda}_p = \mathrm{diag}(\lambda_1, \ldots, \lambda_p).$$

Note that
$$\mathbf{A}_p := [\boldsymbol{a}_1, \ldots, \boldsymbol{a}_p] := [\mathbf{X}_p\boldsymbol{\gamma}_1, \ldots, \mathbf{X}_p\boldsymbol{\gamma}_p] = \mathbf{X}_p\boldsymbol{\Gamma}_p$$
and that the column vectors $\boldsymbol{a}_1, \ldots, \boldsymbol{a}_p$ are called the *principal components* of the vectors of the explanatory variables $\boldsymbol{z}_1, \ldots, \boldsymbol{z}_p$ in the usual model (M_Z).

1.4 Least Squares Estimators

In the model (M_Z), the unknown parameter vector $\boldsymbol{\theta}$ is usually estimated by the least squares method:
$$\Delta^2(\boldsymbol{\theta}) := \|\mathbf{Y} - \mathbf{Z}\boldsymbol{\theta}\|^2 = \|\boldsymbol{\varepsilon}\|^2 \longrightarrow \min_{\boldsymbol{\theta} \in \mathbb{R}^{p+1}},$$

where $\|\boldsymbol{x}\| := \sqrt{\boldsymbol{x}'\boldsymbol{x}} = (x_1^2 + \cdots + x_n^2)^{\frac{1}{2}}$ is the Euclidian norm of $\boldsymbol{x} = [x_1, \cdots, x_n]'$. Differentiating $\Delta^2(\boldsymbol{\theta})$ with respect to $\boldsymbol{\theta}$, the following equation in $\boldsymbol{\theta}$ is obtained:
$$\mathbf{Z}'\mathbf{Z}\boldsymbol{\theta} = \mathbf{Z}'\mathbf{Y}. \tag{1.9}$$

This equation is called the *normal equation*, and its solution
$$\widehat{\boldsymbol{\theta}} := (\mathbf{Z}'\mathbf{Z})^{-1}\mathbf{Z}'\mathbf{Y} \tag{1.10}$$

is called the *ordinary least squares* (OLS) estimator.

The normal equation for $\boldsymbol{\beta}$ in the model (M_X) is
$$\mathbf{X}'\mathbf{X}\boldsymbol{\beta} = \mathbf{X}'\mathbf{Y}, \tag{1.11}$$

and the OLS estimator for β is

$$\widehat{\beta} := (\mathbf{X}'\mathbf{X})^{-1}\mathbf{X}'\mathbf{Y}. \tag{1.12}$$

Similarly, the OLS estimator for α in the model (M_A) is

$$\widehat{\alpha} := (\mathbf{A}'\mathbf{A})^{-1}\mathbf{A}'\mathbf{Y} = \mathbf{\Lambda}^{-1}\mathbf{A}'\mathbf{Y}. \tag{1.13}$$

The variance of the error σ^2 is estimated by

$$\widehat{\sigma}^2 := \frac{1}{n-p-1}(\mathbf{Y} - \mathbf{Z}\widehat{\boldsymbol{\theta}})'(\mathbf{Y} - \mathbf{Z}\widehat{\boldsymbol{\theta}}). \tag{1.14}$$

Note that the estimator is unbiased:

$$\mathrm{E}\left(\widehat{\sigma}^2\right) = \sigma^2.$$

1.5 Invariance of Ordinary Least Squares Estimator

The parameter vectors $\boldsymbol{\theta}$, $\boldsymbol{\beta}$ have the relationship

$$\boldsymbol{\theta} = \mathbf{T}^{-1}\boldsymbol{\beta} \tag{1.15}$$

from (1.5), and their OLS estimators have the similar relationship

$$\widehat{\boldsymbol{\theta}} = \mathbf{T}^{-1}\widehat{\boldsymbol{\beta}}. \tag{1.16}$$

This property is called *invariance* with respect to the transformation of standardization $\mathbf{T}^{-1}$.

Similarly, there exists the following relationship between $\boldsymbol{\beta}$ and $\boldsymbol{\alpha}$ from (1.7):

$$\boldsymbol{\beta} = \boldsymbol{\Gamma}\boldsymbol{\alpha}, \tag{1.17}$$

and the relationship holds also for their OLS estimators:

$$\widehat{\boldsymbol{\beta}} = \boldsymbol{\Gamma}\widehat{\boldsymbol{\alpha}}. \tag{1.18}$$

Thus, the OLS estimator also has invariance with respect to the orthogonal transformation $\boldsymbol{\Gamma}$.

We represent the relationships among the vectors of regression coefficients in the models (M_Z), (M_X), and (M_A) in Figure 1.1 and the relationships among their OLS estimators in Figure 1.2.

$$
\begin{array}{ccccc}
(\mathrm{M}_Z) & & (\mathrm{M}_X) & & (\mathrm{M}_A) \\
& \mathbf{T} & & \boldsymbol{\Gamma}' & \\
\boldsymbol{\theta} & \rightleftarrows & \boldsymbol{\beta} & \rightleftarrows & \boldsymbol{\alpha} \\
& \mathbf{T}^{-1} & & \boldsymbol{\Gamma} &
\end{array}
$$

Figure 1.1: Relationships among the vectors of regression coefficients.

$$
\begin{array}{ccccc}
(\mathrm{M}_Z) & & (\mathrm{M}_X) & & (\mathrm{M}_A) \\
& \mathbf{T} & & \boldsymbol{\Gamma}' & \\
\widehat{\boldsymbol{\theta}} & \rightleftarrows & \widehat{\boldsymbol{\beta}} & \rightleftarrows & \widehat{\boldsymbol{\alpha}} \\
& \mathbf{T}^{-1} & & \boldsymbol{\Gamma} &
\end{array}
$$

Figure 1.2: Relationships among the OLS estimators.

1.6 Multicollinearity

The term *multicollinearity* (or *collinearity*) means that the vectors of the explanatory variables are *nearly linearly dependent*. There have been many reports on this (e.g. Belsley *et al.* (1980) and Gunst (1983)), and they use the model (M_X), which is defined as follows:

Definition 1.1 (Multicollinearity). Let a linear regression model be defined as in (M_X). If there exists a vector $\boldsymbol{\xi}_p := [\xi_1, \ldots, \xi_p]'$ ($\|\boldsymbol{\xi}_p\| = 1$) such that

$$\mathbf{X}_p \boldsymbol{\xi}_p = \xi_1 \boldsymbol{x}_1 + \cdots + \xi_p \boldsymbol{x}_p \simeq \mathbf{0}, \qquad (1.19)$$

then we say multicollinearity exists among the vectors of the explanatory variables $\boldsymbol{x}_1, \ldots, \boldsymbol{x}_p$.

Let us give an important example. If there exist small eigenvalues $\lambda_{r+1} \geq \cdots \geq \lambda_p (> 0)$ of the information matrix $\mathbf{X}_p' \mathbf{X}_p$, then there exist the following multicollinearities with respect to the eigenvector $\boldsymbol{\gamma}_i$ correspondent to the eigenvalue $\lambda_i = \boldsymbol{\gamma}_i' \mathbf{X}_p' \mathbf{X}_p \boldsymbol{\gamma}_i (\simeq 0)$ for all $i \in N_{p \backslash r} := \{r+1, \ldots, p\}$:

$$\mathbf{X}_p \boldsymbol{\gamma}_i \simeq \mathbf{0}, \quad i \in N_{p \backslash r}.$$

In this case, the *total variance* of the OLS estimators $\{\widehat{\beta}_1, \ldots, \widehat{\beta}_p\}$ for $\{\beta_1, \ldots, \beta_p\}$

in the model (M_X) is large:

$$\sum_{i=1}^{p} V(\widehat{\beta}_i) = trace V(\widehat{\boldsymbol{\beta}}_p)$$
$$= \sigma^2 \, trace \, (\mathbf{X}_p'\mathbf{X}_p)^{-1} = \sigma^2 \, trace \boldsymbol{\Lambda}_p^{-1}$$
$$= \sigma^2 \sum_{i=1}^{p} \frac{1}{\lambda_i} = \sigma^2 \sum_{i=1}^{r} \frac{1}{\lambda_i} + \sigma^2 \underbrace{\sum_{i=r+1}^{p} \frac{1}{\lambda_i}}_{\text{large}},$$

where $\widehat{\boldsymbol{\beta}}_p := [\widehat{\beta}_1, \ldots, \widehat{\beta}_p]' := (\mathbf{X}_p'\mathbf{X}_p)^{-1}\mathbf{X}_p'\mathbf{Y}$ and $trace \, \mathbf{A} := a_{11} + \cdots + a_{nn}$ is the trace of a square matrix $\mathbf{A} = [a_{ij}]$ of order n.

More directly, the variance of the OLS estimator $\widehat{\beta}_i$ is

$$V(\widehat{\beta}_i) = \sigma^2 \boldsymbol{\gamma}^i \boldsymbol{\Lambda}_p^{-1} (\boldsymbol{\gamma}^i)' = \sigma^2 \sum_{j=1}^{p} \frac{\gamma_{ij}^2}{\lambda_j},$$

where $\boldsymbol{\gamma}^i$ is the ith row of the orthogonal matrix $\boldsymbol{\Gamma}_p$. Since the rows of $\boldsymbol{\Gamma}_p$ are orthonormal, we must have $\sum_{j=1}^{p} \gamma_{ij}^2 = 1$, so that $|\gamma_{ij}| \leq 1$ for all i and j. Thus, if some eigenvalues λ_i are close to zero and the element γ_{ij} is not close to zero, then $V(\widehat{\beta}_i)$ must be large.

The following index is called the ith *variance inflation factor* (VIF):

$$\text{VIF}_i := \frac{V(\widehat{\beta}_i)}{\sigma^2} = \sum_{j=1}^{p} \frac{\gamma_{ij}^2}{\lambda_j}. \tag{1.20}$$

If there are some large VIFs, then it is possible for variance divergence of the OLS estimator $\widehat{\beta}_i$ to occur. Note that these can be represented as follows:

$$\text{VIF}_i = \frac{1}{1 - R_i^2},$$

where R_i^2 is the *coefficient of determination* when the ith vector of the explanatory variables $\boldsymbol{x}_i$ is regressed to the space spanned by the other vectors of the explanatory variables $\{\boldsymbol{x}_1, \cdots, \boldsymbol{x}_{i-1}, \boldsymbol{x}_{i+1}, \cdots, \boldsymbol{x}_p\}$,

$$R_i^2 := \boldsymbol{x}_i' \mathbf{P}_{(i)} \boldsymbol{x}_i.$$

Note that $\mathbf{X}_{(i)} := [\boldsymbol{x}_1, \cdots, \boldsymbol{x}_{i-1}, \boldsymbol{x}_{i+1}, \cdots, \boldsymbol{x}_p]$ and that

$$\mathbf{P}_{(i)} := \mathbf{X}_{(i)} (\mathbf{X}_{(i)}' \mathbf{X}_{(i)})^{-1} \mathbf{X}_{(i)}$$

is the projection matrix and represents the orthogonal projection onto the linear subspace

$$\mathcal{C}(\mathbf{X}_{(i)}) := \left\{ \mathbf{X}_{(i)} \boldsymbol{\xi}_{(i)} \mid \boldsymbol{\xi}_{(i)} \in \mathbb{R}^{p-1} \right\} (\subset \mathbb{R}^n),$$

which is the column space of $\mathbf{X}_{(i)}$, and for which $\boldsymbol{\xi}_{(i)} := [\xi_1, \ldots, \xi_{i-1}, \xi_{i+1}, \ldots, \xi_p]'$. (See also Section 3.1 in Belsley *et al.* (1980) or Section 9.7 in Seber and Lee (2003).)

Furthermore, the variance of the OLS estimator $\widehat{\alpha}_i$ ($i \in N_{p \backslash r}$) for α_i in the canonical form (M_A) is large:

$$\mathrm{V}(\widehat{\alpha}_i) = \frac{\sigma^2}{\lambda_i}, \ i \in N_{p \backslash r},$$

where $\widehat{\boldsymbol{\alpha}}_p := [\widehat{\alpha}_1, \ldots, \widehat{\alpha}_p]' := (\mathbf{A}_p' \mathbf{A}_p)^{-1} \mathbf{A}_p' \mathbf{Y}$.

The above suggest that if there exist such multicollinearities among the vectors of the explanatory variables, then several problems with respect to the estimating accuracy of the OLS estimator occur.

1.7 Alternative Regression Estimators

In order to handle the problem of variance divergence caused by multicollinearity, the following two devices for solving the normal equation

$$\mathbf{X}_p' \mathbf{X}_p \boldsymbol{\beta}_p = \mathbf{X}_p' \mathbf{Y} \tag{1.21}$$

for $\boldsymbol{\beta}_p$ in the standardized model (M_X) are available, although note that these cause their estimators to be biased:

(D1) Addition of a dummy matrix to $\mathbf{X}_p' \mathbf{X}_p$ ensures the stability of the solution and its variance.

(D2) Reduction of the $p \times p$ information matrix $\mathbf{X}_p' \mathbf{X}_p$ to an "essentially" $r \times r$ matrix ($r < p$) eliminates multicollinearity.

From the point of view of (D1), Hoerl and Kennard (1970a) proposed the *ordinary ridge regression* (ORR) estimator and the *generalized ridge regression* (GRR) estimator. From the point of view of (D2), Kendall (1957) proposed and Marquardt (1970) investigated the *principal component regression* (PCR) estimator where r is the number of principal components. (Jolliffe (2002) is a detailed treatment of the whole of principal component analysis.) Furthermore, we have the *r-k class* estimator derived from a combination of the above two devices, which was suggested by Marquardt (1970) and examined in detail by Baye and Parker (1984). Note that all of these are referred to as *shrinkage estimators* or *shrinkage regression estimators*. For example, the ORR estimator $\widehat{\boldsymbol{\beta}}_p(\cdot, k)$ and the PCR estimator $\widehat{\boldsymbol{\beta}}_p(r, \cdot)$ shrink the OLS estimator $\widehat{\boldsymbol{\beta}}_p$ to the origin as follows:

$$\|\widehat{\boldsymbol{\beta}}_p(\cdot, k)\| < \|\widehat{\boldsymbol{\beta}}_p\|, \quad k > 0,$$
$$\|\widehat{\boldsymbol{\beta}}_p(r, \cdot)\| < \|\widehat{\boldsymbol{\beta}}_p\|, \quad r < p.$$

In the next chapter, we give the definitions of all of these estimators.

Chapter 2
Shrinkage Regression Estimators

In this chapter, we consider several shrinkage regression estimators. We also mention several shrinkage-type regression estimators for the regression coefficients in the usual regression model.

2.1 Ordinary Ridge Regression Estimator

Hoerl and Kennard (1970a) considered stabilizing the solution of the normal equation and treated the following equation obtained by adding a small non-negative quantity k to the diagonal elements of the correlation form $\mathbf{X}_p'\mathbf{X}_p$ in (1.21):

$$(\mathbf{X}_p'\mathbf{X}_p + k\mathbf{I}_p)\boldsymbol{\beta}_p = \mathbf{X}_p'\boldsymbol{Y}, \tag{2.1}$$

where $k(\geq 0)$ is called a *ridge coefficient*. Note that this idea of addition is based on device (D1) in Section 1.7. By solving (2.1) with respect to $\boldsymbol{\beta}_p$, we have

$$\widehat{\boldsymbol{\beta}}_p(\cdot, k) := (\mathbf{X}_p'\mathbf{X}_p + k\mathbf{I}_p)^{-1}\mathbf{X}_p'\boldsymbol{Y}. \tag{2.2}$$

The solution $\widehat{\boldsymbol{\beta}}_p(\cdot, k)$ is called the *ordinary ridge regression* (ORR) estimator for $\boldsymbol{\beta}_p$. Note that if $k = 0$, then the ORR estimator is equal to the OLS estimator:

$$\widehat{\boldsymbol{\beta}}_p(\cdot, 0) = (\mathbf{X}_p'\mathbf{X}_p)^{-1}\mathbf{X}_p'\boldsymbol{Y} = \widehat{\boldsymbol{\beta}}_p.$$

2.2 Generalized Ridge Regression Estimator

Hoerl and Kennard (1970a) also proposed a generalized version of the ORR estimator $\widehat{\boldsymbol{\beta}}_p(\cdot, k)$, which is defined by the following procedure. First of all, we consider the following equation obtained by adding the dummy diagonal matrix $\mathbf{K}_p := \mathrm{diag}(k_1, \ldots, k_p)$ to $\mathbf{A}_p'\mathbf{A}_p$:

$$(\mathbf{A}_p'\mathbf{A}_p + \mathbf{K}_p)\boldsymbol{\alpha}_p = \mathbf{A}_p'\boldsymbol{Y}, \tag{2.3}$$

where $k_i(\geq 0)$ is called the ith ridge coefficient. Next, by solving (2.3) with respect to $\boldsymbol{\alpha}_p$, we have

$$\widehat{\boldsymbol{\alpha}}_p(\mathbf{K}_p) := (\mathbf{A}_p'\mathbf{A}_p + \mathbf{K}_p)^{-1}\mathbf{A}_p'\boldsymbol{Y}. \tag{2.4}$$

The solution $\widehat{\boldsymbol{\alpha}}_p(\mathbf{K}_p) = [\widehat{\alpha}_1(k_1), \ldots, \widehat{\alpha}_p(k_p)]'$ is called the *generalized ridge regression* (GRR) estimator for $\boldsymbol{\alpha}_p$. Finally, by using the relationship $\boldsymbol{\beta}_p = \boldsymbol{\Gamma}_p\boldsymbol{\alpha}_p$, we can define the GRR estimator for $\boldsymbol{\beta}_p$ as follows:

$$\widehat{\boldsymbol{\beta}}_p(\mathbf{K}_p) := \boldsymbol{\Gamma}_p\widehat{\boldsymbol{\alpha}}_p(\mathbf{K}_p) = (\mathbf{X}_p'\mathbf{X}_p + \boldsymbol{\Gamma}_p\mathbf{K}_p\boldsymbol{\Gamma}_p')^{-1}\mathbf{X}_p'\boldsymbol{Y}. \tag{2.5}$$

(See also Figure 1.1.) Note that if $k_1 = \cdots = k_p = k$, then the GRR estimator is equal to the ORR estimator, and thus if $k_1 = \cdots = k_p = 0$, then the GRR estimator is equal to the OLS estimator:

$$\widehat{\boldsymbol{\beta}}_p(k\mathbf{I}_p) = (\mathbf{X}'_p\mathbf{X}_p + k\mathbf{I}_p)^{-1}\mathbf{X}'_p\boldsymbol{Y} = \widehat{\boldsymbol{\beta}}_p(\cdot, k),$$
$$\widehat{\boldsymbol{\beta}}_p(\mathbf{O}) = (\mathbf{X}'_p\mathbf{X}_p)^{-1}\mathbf{X}'_p\boldsymbol{Y} = \widehat{\boldsymbol{\beta}}_p.$$

2.3 Principal Component Regression Estimator

Kendall (1957) applied principal component analysis to regression analysis. The *principal component regression* (PCR) estimator is defined by the following procedure. First of all, the matrix of all principal components $\mathbf{A}_p := \mathbf{X}_p\boldsymbol{\Gamma}_p$ in the model (M_A) is reduced to a matrix of some of the principal components, $\mathbf{A}_r$ ($r < p$), where $\mathbf{A}_p = [\mathbf{A}_r, \mathbf{A}_{p\setminus r}]$, and $\mathbf{A}_{p\setminus r}$ is the matrix of the principal components corresponding to small eigenvalues $\lambda_{r+1}, \ldots, \lambda_p$ of the information matrix $\mathbf{X}'_p\mathbf{X}_p$. Note that this idea of reduction is based on device (D2) in Section 1.7. Next, we regress (project) the vector of response variables $\boldsymbol{Y}$ to the column space $\mathcal{C}(\mathbf{A}_r)$ spanned by the principal components $\{\boldsymbol{a}_1, \ldots, \boldsymbol{a}_r\}$, which gives the following vector as the estimated regression coefficients:

$$\widehat{\boldsymbol{\alpha}}_r := (\mathbf{A}'_r\mathbf{A}_r)^{-1}\mathbf{A}'_r\boldsymbol{Y} = \boldsymbol{\Lambda}_r^{-1}\mathbf{A}'_r\boldsymbol{Y}.$$

Finally, by using the orthogonal transformation $\boldsymbol{\Gamma}_r$, we return to the model (M_X) as follows:

$$\widehat{\boldsymbol{\beta}}_p(r, \cdot) := \boldsymbol{\Gamma}_r\widehat{\boldsymbol{\alpha}}_r = \boldsymbol{\Gamma}_r\boldsymbol{\Lambda}_r^{-1}\boldsymbol{\Gamma}'_r\mathbf{X}'_p\boldsymbol{Y}. \tag{2.6}$$

This estimator $\widehat{\boldsymbol{\beta}}_p(r, \cdot)$ is the PCR estimator for $\boldsymbol{\beta}_p$. Note that if $r = p$, then the PCR estimator is equal to the OLS estimator:

$$\widehat{\boldsymbol{\beta}}_p(p, \cdot) = \boldsymbol{\Gamma}_p\boldsymbol{\Lambda}_p^{-1}\boldsymbol{\Gamma}'_p\mathbf{X}'_p\boldsymbol{Y} = \widehat{\boldsymbol{\beta}}_p.$$

Remark 2.1. *If we define*

$$(\mathbf{X}'_p\mathbf{X}_p)^- := \boldsymbol{\Gamma}_r\boldsymbol{\Lambda}_r^{-1}\boldsymbol{\Gamma}'_r,$$

then we know

$$(\mathbf{X}'_p\mathbf{X}_p)(\mathbf{X}'_p\mathbf{X}_p)^-(\mathbf{X}'_p\mathbf{X}_p) = (\mathbf{X}'_p\mathbf{X}_p), \quad (\mathbf{X}'_p\mathbf{X}_p)^-(\mathbf{X}'_p\mathbf{X}_p)(\mathbf{X}'_p\mathbf{X}_p)^- = (\mathbf{X}'_p\mathbf{X}_p)^-.$$

Hence, $(\mathbf{X}'_p\mathbf{X}_p)^-$ is a generalized inverse (g-inverse) matrix. (Strictly speaking, the matrix $(\mathbf{X}'_p\mathbf{X}_p)^-$ is the Moore-Penrose-type g-inverse matrix. See Rao (1973) for the details of the g-inverse matrix.) Then,

$$\widehat{\boldsymbol{\beta}}_p^- := (\mathbf{X}'_p\mathbf{X}_p)^-\mathbf{X}'_p\boldsymbol{Y}$$

is called the generalized inverse regression *estimator and satisfies the normal equation* (1.21) *when rank* $(\mathbf{X}'_p\mathbf{X}_p) = r(<p)(; singular)$. *Marquardt* (1970) *investigated a class of biased linear estimators including the g-inverse estimator and the ORR estimator in detail. Note that the g-inverse estimator is equal to the PCR estimator:*

$$\widehat{\boldsymbol{\beta}}_p^- = (\mathbf{X}'_p\mathbf{X}_p)^-\mathbf{X}'_p\boldsymbol{Y} = \boldsymbol{\Gamma}_r\boldsymbol{\Lambda}_r^{-1}\boldsymbol{\Gamma}'_r\mathbf{X}'_p\boldsymbol{Y} = \widehat{\boldsymbol{\beta}}_p(r,\cdot).$$

2.4 r-k Class Estimator

Baye and Parker (1984) attempted a compromise between the PCR estimator and the ORR estimators, proposing the *r-k class* estimator as an analogy of the OLS estimator versus the ORR estimators. Note that this estimator is based on both devices (D1) and (D2) in Section 1.7.

The r-k class estimator is defined as follows:

$$\widehat{\boldsymbol{\beta}}_p(r,k) := \boldsymbol{\Gamma}_r\left(\boldsymbol{\Lambda}_r + k\mathbf{I}_p\right)^{-1}\boldsymbol{\Gamma}'_r\mathbf{X}'_p\boldsymbol{Y}, \qquad (2.7)$$

where r is a number of principal components and k is the ridge coefficient. Note that the OLS, PCR, and ORR estimators are rewritten using this notation as

$$\widehat{\boldsymbol{\beta}}_p = \widehat{\boldsymbol{\beta}}_p(p,0), \quad \widehat{\boldsymbol{\beta}}_p(r,\cdot) = \widehat{\boldsymbol{\beta}}_p(r,0), \quad \widehat{\boldsymbol{\beta}}_p(\cdot,k) = \widehat{\boldsymbol{\beta}}_p(p,k), \qquad (2.8)$$

respectively.

2.5   Invariance of Shrinkage Regression Estimators

The shrinkage regression estimators have been considered under the standardized model (M_X), and are used to estimate the vector of the regression coefficients $\boldsymbol{\beta}_p$. However, simultaneously performing shrinkage estimation for β_0 is also important, as is that for the $\boldsymbol{\theta}$ in the usual model (M_Z).

Brown (1977) discussed simultaneous estimation for $[\beta_0, \boldsymbol{\beta}'_p]'(=\boldsymbol{\beta})$ by the following (usual) estimator:

$$\widehat{\boldsymbol{\beta}}(\cdot,0) := \begin{bmatrix} \widehat{\beta}_0(0) \\ \widehat{\boldsymbol{\beta}}_p(\cdot,k) \end{bmatrix} := \begin{bmatrix} \overline{Y} \\ (\mathbf{X}'_p\mathbf{X}_p + k\mathbf{I}_p)^{-1}\mathbf{X}'_p\boldsymbol{Y} \end{bmatrix},$$

where $\overline{Y} = \sum_{i=1}^n Y_i/n$, $k \geq 0$, and pointed out that this estimator has location invariance. (See Theorem in Brown (1977).) The first coordinate, $\widehat{\beta}_0(0) := \overline{Y} =: \widehat{\beta}_0$, is the OLS estimator for the intercept β_0, and it is not shrunk in the same way as the residual part, $\widehat{\boldsymbol{\beta}}_p(\cdot,k)$, which is the ORR estimator for $\boldsymbol{\beta}_p$. The estimator may be written using the notation of generalized ridge regression as

$$\widehat{\boldsymbol{\beta}}(\cdot,0) = (\mathbf{X}'\mathbf{X} + k\,\mathrm{diag}(0,1,..,1))^{-1}\mathbf{X}'\boldsymbol{Y}, \qquad (2.9)$$

where diag(0,1,..,1) is the diagonal matrix of order $(p+1)$ with the first diagonal element 0 and the others 1.

Jimichi and Inagaki (1993) considered the slightly more general ridge estimation for β, and proposed the following estimator:

$$\widehat{\beta}(\cdot, k_0) := \begin{bmatrix} \widehat{\beta}_0(k_0) \\ \widehat{\beta}_p(\cdot, k) \end{bmatrix} := (\mathbf{X}'\mathbf{X} + \mathrm{diag}(k_0, k, \ldots, k))^{-1}\mathbf{X}'\mathbf{Y} \qquad (2.10)$$

$$= \begin{bmatrix} \dfrac{n}{n+k_0}\overline{Y} \\ (\mathbf{X}'_p\mathbf{X}_p + k\mathbf{I}_p)^{-1}\mathbf{X}'_p\mathbf{Y} \end{bmatrix}, \quad k_0 > 0.$$

Note that the estimator

$$\widehat{\beta}(\cdot, k) := \begin{bmatrix} \widehat{\beta}_0(k) \\ \widehat{\beta}_p(\cdot, k) \end{bmatrix} := (\mathbf{X}'\mathbf{X} + k\mathbf{I}_{p+1})^{-1}\mathbf{X}'\mathbf{Y} \qquad (2.11)$$

$$= \begin{bmatrix} \dfrac{n}{n+k}\overline{Y} \\ (\mathbf{X}'_p\mathbf{X}_p + k\mathbf{I}_p)^{-1}\mathbf{X}'_p\mathbf{Y} \end{bmatrix}$$

is the ORR estimator for β and the estimator $\widehat{\beta}(\cdot, k_0)$ includes the estimators $\widehat{\beta}(\cdot, 0)$, $\widehat{\beta}(\cdot, k)$ as special cases:

$$\widehat{\beta}(\cdot, k_0) = \begin{cases} \widehat{\beta}(\cdot, 0), & k_0 = 0, \\ \widehat{\beta}(\cdot, k), & k_0 = k. \end{cases}$$

Now, let us consider ridge estimation for $\boldsymbol{\theta}$ in the model (M_Z). Recall the following equivalent equation (1.15):

$$\boldsymbol{\theta} = \mathbf{T}^{-1}\boldsymbol{\beta} \iff \begin{bmatrix} \theta_0 \\ \boldsymbol{\theta}_p \end{bmatrix} = \begin{bmatrix} \beta_0 - \boldsymbol{m}'_p\mathbf{S}_p^{-1}\boldsymbol{\beta}_p \\ \mathbf{S}_p^{-1}\boldsymbol{\beta}_p \end{bmatrix}. \qquad (2.12)$$

(See also Figure 1.1.) Let us call the following estimators for $\boldsymbol{\theta}$ obtained by substituting the previous ridge estimators for $\boldsymbol{\beta}$ into (2.12) the *ORR-type* estimators for $\boldsymbol{\theta}$:

$$\widehat{\boldsymbol{\theta}}(\cdot, 0) := \mathbf{T}^{-1}\widehat{\boldsymbol{\beta}}(\cdot, 0) = \begin{bmatrix} \widehat{\beta}_0(0) - \boldsymbol{m}'_p\mathbf{S}_p^{-1}\widehat{\boldsymbol{\beta}}_p(\cdot, k) \\ \mathbf{S}_p^{-1}\widehat{\boldsymbol{\beta}}_p(\cdot, k) \end{bmatrix}, \qquad (2.13)$$

$$\widehat{\boldsymbol{\theta}}(\cdot, k_0) := \mathbf{T}^{-1}\widehat{\boldsymbol{\beta}}(\cdot, k_0) = \begin{bmatrix} \widehat{\beta}_0(k_0) - \boldsymbol{m}'_p\mathbf{S}_p^{-1}\widehat{\boldsymbol{\beta}}_p(\cdot, k) \\ \mathbf{S}_p^{-1}\widehat{\boldsymbol{\beta}}_p(\cdot, k) \end{bmatrix}, \qquad (2.14)$$

$$\widehat{\boldsymbol{\theta}}(\cdot, k) := \mathbf{T}^{-1}\widehat{\boldsymbol{\beta}}(\cdot, k) = \begin{bmatrix} \widehat{\beta}_0(k) - \boldsymbol{m}'_p\mathbf{S}_p^{-1}\widehat{\boldsymbol{\beta}}_p(\cdot, k) \\ \mathbf{S}_p^{-1}\widehat{\boldsymbol{\beta}}_p(\cdot, k) \end{bmatrix}. \qquad (2.15)$$

Note that the first ridge-type estimator (2.13) is often used for estimating $\boldsymbol{\theta}$ (e.g. Draper and Smith (1981)).

2.5. INVARIANCE OF SHRINKAGE REGRESSION ESTIMATORS

Remark 2.2. *Although the shrinkage regression estimators $\widehat{\boldsymbol{\beta}}(\cdot, k_0)$, $\widehat{\boldsymbol{\beta}}(\cdot, 0)$ for $\boldsymbol{\beta}$ and the shrinkage-type regression estimators $\widehat{\boldsymbol{\theta}}(\cdot, k_0)$, $\widehat{\boldsymbol{\theta}}(\cdot, 0)$ for $\boldsymbol{\theta}$ depend on the ridge coefficient k, we abbreviate the notation for simplicity.*

Remark 2.3. *The OLS estimator is invariant. That is, the OLS estimator $\widehat{\boldsymbol{\theta}}$ for $\boldsymbol{\theta}$ can be obtained by substituting the OLS estimator $\widehat{\boldsymbol{\beta}}$ for $\boldsymbol{\beta}$ into (2.12):*

$$\widehat{\boldsymbol{\theta}} = (\mathbf{Z}'\mathbf{Z})^{-1}\mathbf{Z}'\mathbf{Y} = \mathbf{T}^{-1}\widehat{\boldsymbol{\beta}}.$$

However, unfortunately, none of the shrinkage regression estimators considered in this book have any invariances. For example, the estimator $\widehat{\boldsymbol{\theta}}(\cdot, k) = \mathbf{T}^{-1}\widehat{\boldsymbol{\beta}}(\cdot, k)$ is not equal to the ORR estimator $\widetilde{\boldsymbol{\theta}}(\cdot, k)$ for $\boldsymbol{\theta}$:

$$\widetilde{\boldsymbol{\theta}}(\cdot, k) := (\mathbf{Z}'\mathbf{Z} + k\mathbf{I}_{p+1})^{-1}\mathbf{Z}'\mathbf{Y} \neq \widehat{\boldsymbol{\theta}}(\cdot, k).$$

As a consequence, if we want to estimate $\boldsymbol{\theta}$ in the usual model (M_Z) by using shrinkage estimators, then we must transform the estimators for $\boldsymbol{\beta}$ by $\mathbf{T}^{-1}$. See also Jimichi and Inagaki (1993).

We can consider the following GRR estimators for $\boldsymbol{\beta}$:

$$\widehat{\boldsymbol{\beta}}(0, \mathbf{K}_p) := \begin{bmatrix} \widehat{\beta}_0(0) \\ \widehat{\boldsymbol{\beta}}_p(\mathbf{K}_p) \end{bmatrix} = \begin{bmatrix} \overline{Y} \\ (\mathbf{X}_p'\mathbf{X}_p + \boldsymbol{\Gamma}_p\mathbf{K}_p\boldsymbol{\Gamma}_p')^{-1}\mathbf{X}_p'\mathbf{Y} \end{bmatrix}, \quad (2.16)$$

$$\widehat{\boldsymbol{\beta}}(k_0, \mathbf{K}_p) := \begin{bmatrix} \widehat{\beta}_0(k_0) \\ \widehat{\boldsymbol{\beta}}_p(\mathbf{K}_p) \end{bmatrix} = \begin{bmatrix} \frac{n}{n+k_0}\overline{Y} \\ (\mathbf{X}_p'\mathbf{X}_p + \boldsymbol{\Gamma}_p\mathbf{K}_p\boldsymbol{\Gamma}_p')^{-1}\mathbf{X}_p'\mathbf{Y} \end{bmatrix}. \quad (2.17)$$

Thus, from Remark 2.3, we can obtain the estimators for $\boldsymbol{\theta}$ as follows:

$$\widehat{\boldsymbol{\theta}}(0, \mathbf{K}_p) := \mathbf{T}^{-1}\widehat{\boldsymbol{\beta}}(0, \mathbf{K}_p) = \begin{bmatrix} \widehat{\beta}_0(0) - \boldsymbol{m}_p'\mathbf{S}_p^{-1}\widehat{\boldsymbol{\beta}}_p(\mathbf{K}_p) \\ \mathbf{S}_p^{-1}\widehat{\boldsymbol{\beta}}_p(\mathbf{K}_p) \end{bmatrix}, \quad (2.18)$$

$$\widehat{\boldsymbol{\theta}}(k_0, \mathbf{K}_p) := \mathbf{T}^{-1}\widehat{\boldsymbol{\beta}}(k_0, \mathbf{K}_p) = \begin{bmatrix} \widehat{\beta}_0(k_0) - \boldsymbol{m}_p'\mathbf{S}_p^{-1}\widehat{\boldsymbol{\beta}}_p(\mathbf{K}_p) \\ \mathbf{S}_p^{-1}\widehat{\boldsymbol{\beta}}_p(\mathbf{K}_p) \end{bmatrix}. \quad (2.19)$$

Let us call these estimators the *GRR-type* estimators.

If we use the PCR estimator $\widehat{\boldsymbol{\beta}}_p(r, \cdot)$ for $\boldsymbol{\beta}_p$, then an ordinary estimator for $\boldsymbol{\beta}$ is

$$\widehat{\boldsymbol{\beta}}(r, \cdot) := \begin{bmatrix} \widehat{\beta}_0 \\ \widehat{\boldsymbol{\beta}}_p(r, \cdot) \end{bmatrix} := \begin{bmatrix} \overline{Y} \\ \boldsymbol{\Gamma}_r\boldsymbol{\Lambda}_r^{-1}\boldsymbol{\Gamma}_r'\mathbf{X}_p'\mathbf{Y} \end{bmatrix},$$

and the estimator for $\boldsymbol{\theta}$ is obtained as follows:

$$\widehat{\boldsymbol{\theta}}(r, \cdot) := \mathbf{T}^{-1}\widehat{\boldsymbol{\beta}}(r, \cdot) = \begin{bmatrix} \widehat{\beta}_0 - \boldsymbol{m}_p'\mathbf{S}_p^{-1}\widehat{\boldsymbol{\beta}}_p(r, \cdot) \\ \mathbf{S}_p^{-1}\widehat{\boldsymbol{\beta}}_p(r, \cdot) \end{bmatrix}. \quad (2.20)$$

Let us call this estimator the *PCR-type* estimator.

Finally, we can consider the following r-k class estimations for $\boldsymbol{\beta}$:

$$\widehat{\boldsymbol{\beta}}(r,0) := \begin{bmatrix} \widehat{\beta}_0(0) \\ \widehat{\boldsymbol{\beta}}_p(r,k) \end{bmatrix} = \begin{bmatrix} \overline{Y} \\ \boldsymbol{\Gamma}_r \left(\boldsymbol{\Lambda}_r + k \mathbf{I}_p \right)^{-1} \boldsymbol{\Gamma}_r' \mathbf{X}_p' \boldsymbol{Y} \end{bmatrix}, \qquad (2.21)$$

$$\widehat{\boldsymbol{\beta}}(r,k_0) := \begin{bmatrix} \widehat{\beta}_0(k_0) \\ \widehat{\boldsymbol{\beta}}_p(r,k) \end{bmatrix} = \begin{bmatrix} \frac{n}{n+k_0} \overline{Y} \\ \boldsymbol{\Gamma}_r \left(\boldsymbol{\Lambda}_r + k \mathbf{I}_p \right)^{-1} \boldsymbol{\Gamma}_r' \mathbf{X}_p' \boldsymbol{Y} \end{bmatrix}, \qquad (2.22)$$

$$\widehat{\boldsymbol{\beta}}(r,k) := \begin{bmatrix} \widehat{\beta}_0(k) \\ \widehat{\boldsymbol{\beta}}_p(r,k) \end{bmatrix} = \begin{bmatrix} \frac{n}{n+k} \overline{Y} \\ \boldsymbol{\Gamma}_r \left(\boldsymbol{\Lambda}_r + k \mathbf{I}_p \right)^{-1} \boldsymbol{\Gamma}_r' \mathbf{X}_p' \boldsymbol{Y} \end{bmatrix}. \qquad (2.23)$$

Thus, we obtain the following estimators for $\boldsymbol{\theta}$ called *r-k-class-type* estimators:

$$\widehat{\boldsymbol{\theta}}(r,0) := \mathbf{T}^{-1} \widehat{\boldsymbol{\beta}}(r,0) = \begin{bmatrix} \widehat{\beta}_0(0) - \boldsymbol{m}_p' \mathbf{S}_p^{-1} \widehat{\boldsymbol{\beta}}_p(r,k) \\ \mathbf{S}_p^{-1} \widehat{\boldsymbol{\beta}}_p(r,k) \end{bmatrix}, \qquad (2.24)$$

$$\widehat{\boldsymbol{\theta}}(r,k_0) := \mathbf{T}^{-1} \widehat{\boldsymbol{\beta}}(r,k_0) = \begin{bmatrix} \widehat{\beta}_0(k_0) - \boldsymbol{m}_p' \mathbf{S}_p^{-1} \widehat{\boldsymbol{\beta}}_p(r,k) \\ \mathbf{S}_p^{-1} \widehat{\boldsymbol{\beta}}_p(r,k) \end{bmatrix}, \qquad (2.25)$$

$$\widehat{\boldsymbol{\theta}}(r,k) := \mathbf{T}^{-1} \widehat{\boldsymbol{\beta}}(r,k) = \begin{bmatrix} \widehat{\beta}_0(k) - \boldsymbol{m}_p' \mathbf{S}_p^{-1} \widehat{\boldsymbol{\beta}}_p(r,k) \\ \mathbf{S}_p^{-1} \widehat{\boldsymbol{\beta}}_p(r,k) \end{bmatrix}. \qquad (2.26)$$

Remark 2.4. *Although the shrinkage regression estimators $\widehat{\boldsymbol{\beta}}(r,k_0)$, $\widehat{\boldsymbol{\beta}}(r,0)$ for $\boldsymbol{\beta}$ and the shrinkage-type regression estimators $\widehat{\boldsymbol{\theta}}(r,k_0)$, $\widehat{\boldsymbol{\theta}}(r,0)$ for $\boldsymbol{\theta}$ depend on the ridge coefficient k, we abbreviate the notation for simplicity.*

Note that several problems with respect to selecting 0, k, and k_0 in these estimators are treated in Section 3.4. We also consider basic methods for choosing the number of principal components and the ridge coefficients.

Chapter 3
Mean Squared Error Criteria

In this chapter, several mean squared error criteria for the shrinkage regression estimators are given, and their total mean squared errors are compared. Some methods for choosing the number of principal components and ridge coefficients are also considered.

3.1 Definitions of Mean Squared Error Criteria

Let us give the definitions of criteria related to mean squared error in the general case. Assume that $\boldsymbol{\theta}$ is a vector of parameters and its estimator is $\widehat{\boldsymbol{\theta}}$, and let $\mathrm{E}(\widehat{\boldsymbol{\theta}}) := \mathrm{E}_F(\widehat{\boldsymbol{\theta}})$, $\mathrm{V}(\widehat{\boldsymbol{\theta}}) := \mathrm{V}_F(\widehat{\boldsymbol{\theta}})$ denote its expectation and variance, respectively, under the distribution F. The following matrix is called the *mean squared error* (MSE) matrix of $\widehat{\boldsymbol{\theta}}$:

$$\mathrm{MSE}(\widehat{\boldsymbol{\theta}}) := \mathrm{E}(\widehat{\boldsymbol{\theta}} - \boldsymbol{\theta})(\widehat{\boldsymbol{\theta}} - \boldsymbol{\theta})'. \tag{3.1}$$

Note that this can be rewritten as

$$\mathrm{MSE}(\widehat{\boldsymbol{\theta}}) = \mathrm{V}(\widehat{\boldsymbol{\theta}}) + \mathrm{bias}(\widehat{\boldsymbol{\theta}})\mathrm{bias}(\widehat{\boldsymbol{\theta}})', \tag{3.2}$$

where $\mathrm{V}(\widehat{\boldsymbol{\theta}}) := \mathrm{E}(\widehat{\boldsymbol{\theta}} - \mathrm{E}(\widehat{\boldsymbol{\theta}}))(\widehat{\boldsymbol{\theta}} - \mathrm{E}(\widehat{\boldsymbol{\theta}}))'$ is the *variance-covariance matrix* of $\widehat{\boldsymbol{\theta}}$ and $\mathrm{bias}(\widehat{\boldsymbol{\theta}}) := \mathrm{E}(\widehat{\boldsymbol{\theta}}) - \boldsymbol{\theta}$ is the *bias vector*. Its components are

$$[\mathrm{MSE}(\widehat{\boldsymbol{\theta}})]_{ij} = \begin{cases} \mathrm{E}(\widehat{\theta}_i - \theta_i)^2 =: \mathrm{MSE}(\widehat{\theta}_i), & i = j, \\ \mathrm{E}(\widehat{\theta}_i - \theta_i)(\widehat{\theta}_j - \theta_j) =: \mathrm{MCE}(\widehat{\theta}_i, \widehat{\theta}_j), & i \neq j, \end{cases} \tag{3.3}$$

where $\mathrm{MSE}(\widehat{\theta}_i)$ is the MSE of the estimator $\widehat{\theta}_i$, and $\mathrm{MCE}(\widehat{\theta}_i, \theta_j)$ is the *mean cross error* (MCE) between $\widehat{\theta}_i$ and $\widehat{\theta}_j$.

For the shrinkage regression estimators, the most popular criterion is perhaps the *total mean squared error* (TMSE) defined by

$$\mathrm{TMSE}(\widehat{\boldsymbol{\theta}}) := \sum_i \mathrm{MSE}(\widehat{\theta}_i). \tag{3.4}$$

Note that this can be rewritten as

$$\begin{aligned}\mathrm{TMSE}(\widehat{\boldsymbol{\theta}}) &= \textit{trace } \mathrm{MSE}(\widehat{\boldsymbol{\theta}}) \\ &= \mathrm{E}(\widehat{\boldsymbol{\theta}} - \boldsymbol{\theta})'(\widehat{\boldsymbol{\theta}} - \boldsymbol{\theta}) = \mathrm{E}\|\widehat{\boldsymbol{\theta}} - \boldsymbol{\theta}\|^2 \quad (3.5) \\ &= \textit{trace } \mathrm{V}(\widehat{\boldsymbol{\theta}}) + \|\mathrm{bias}(\widehat{\boldsymbol{\theta}})\|^2. \quad (3.6)\end{aligned}$$

Remark 3.1. *The TMSE is a natural criterion from the viewpoint of statistical decision theory. That is, if a quadratic loss function*

$$L(\boldsymbol{\theta}, \widehat{\boldsymbol{\theta}}) := (\boldsymbol{\theta} - \widehat{\boldsymbol{\theta}})'(\boldsymbol{\theta} - \widehat{\boldsymbol{\theta}})$$

is adopted, then the TMSE is equal to the risk function

$$R(\boldsymbol{\theta}, \widehat{\boldsymbol{\theta}}) := E_{\widehat{\boldsymbol{\theta}}} L(\boldsymbol{\theta}, \widehat{\boldsymbol{\theta}}) = TMSE(\widehat{\boldsymbol{\theta}}).$$

3.2 Mean Squared Error Criteria of Regression Estimators

In this section, we consider the MSE criteria of the regression estimators for the vectors of regression coefficients $\boldsymbol{\beta}_p$, $\boldsymbol{\beta}$ and $\boldsymbol{\theta}$.

3.2.1 OLS Estimator

First of all, we will discuss the MSE criteria of the OLS estimator. We start by considering the following OLS estimator for $\boldsymbol{\beta}_p$:

$$\widehat{\boldsymbol{\beta}}_p = (\mathbf{X}_p' \mathbf{X}_p)^{-1} \mathbf{X}_p' \mathbf{Y}.$$

Because the OLS estimator is unbiased,

$$\mathrm{E}(\widehat{\boldsymbol{\beta}}_p) = \boldsymbol{\beta}_p,$$

the bias vector of $\widehat{\boldsymbol{\beta}}_p$ is the zero vector,

$$\mathrm{bias}(\widehat{\boldsymbol{\beta}}_p) = \mathrm{E}(\widehat{\boldsymbol{\beta}}_p) - \boldsymbol{\beta}_p = \mathbf{0}.$$

The variance-covariance matrix of the OLS estimator is

$$\mathrm{V}(\widehat{\boldsymbol{\beta}}_p) = \sigma^2 (\mathbf{X}_p' \mathbf{X}_p)^{-1}.$$

Therefore, from (3.2), the MSE matrix is

$$\begin{aligned}\mathrm{MSE}(\widehat{\boldsymbol{\beta}}_p) &= \mathrm{V}(\widehat{\boldsymbol{\beta}}_p) + \mathrm{bias}(\widehat{\boldsymbol{\beta}}_p)\mathrm{bias}(\widehat{\boldsymbol{\beta}}_p)' \\ &= \sigma^2 (\mathbf{X}_p' \mathbf{X}_p)^{-1},\end{aligned} \qquad (3.7)$$

and the TMSE is

$$\begin{aligned}\mathrm{TMSE}(\widehat{\boldsymbol{\beta}}_p) &= trace\ \mathrm{MSE}(\widehat{\boldsymbol{\beta}}_p) = \sigma^2 trace(\mathbf{X}_p' \mathbf{X}_p)^{-1} \\ &= \sigma^2 trace \boldsymbol{\Gamma}_p \boldsymbol{\Lambda}_p^{-1} \boldsymbol{\Gamma}_p = \sigma^2 trace \boldsymbol{\Lambda}_p^{-1} \\ &= \sigma^2 \sum_{i=1}^{p} \frac{1}{\lambda_i}.\end{aligned} \qquad (3.8)$$

3.2. MEAN SQUARED ERROR CRITERIA OF REGRESSION ESTIMATORS

Next, we consider the MSE criteria of the OLS estimator

$$\widehat{\boldsymbol{\beta}} = (\mathbf{X}'\mathbf{X})^{-1}\mathbf{X}'\boldsymbol{Y}$$
$$= \begin{bmatrix} n & \mathbf{0} \\ \mathbf{0} & \mathbf{X}'_p\mathbf{X}_p \end{bmatrix}^{-1} \begin{bmatrix} \mathbf{1}' \\ \mathbf{X}'_p \end{bmatrix} \boldsymbol{Y} = \begin{bmatrix} \overline{Y} \\ (\mathbf{X}'_p\mathbf{X}_p)^{-1}\mathbf{X}'_p\boldsymbol{Y} \end{bmatrix}$$
$$= \begin{bmatrix} \widehat{\beta}_0 \\ \widehat{\boldsymbol{\beta}}_p \end{bmatrix}.$$

Similar to the case $\widehat{\boldsymbol{\beta}}_p$,

$$\mathrm{E}(\widehat{\boldsymbol{\beta}}) = \boldsymbol{\beta} = \begin{bmatrix} \beta_0 \\ \boldsymbol{\beta}_p \end{bmatrix} \Leftrightarrow \mathrm{bias}(\widehat{\boldsymbol{\beta}}) = \mathbf{0} = \begin{bmatrix} 0 \\ \mathbf{0} \end{bmatrix} = \begin{bmatrix} \mathrm{bias}(\widehat{\beta}_0) \\ \mathrm{bias}(\widehat{\boldsymbol{\beta}}_p) \end{bmatrix},$$

$$\mathrm{V}(\widehat{\boldsymbol{\beta}}) = \sigma^2(\mathbf{X}'\mathbf{X})^{-1} = \begin{bmatrix} \frac{\sigma^2}{n} & \mathbf{0} \\ \mathbf{0} & \sigma^2(\mathbf{X}'_p\mathbf{X}_p)^{-1} \end{bmatrix} = \begin{bmatrix} \mathrm{V}(\widehat{\beta}_0) & \mathrm{Cov}(\widehat{\beta}_0, \widehat{\boldsymbol{\beta}}_p) \\ \mathrm{Cov}(\widehat{\boldsymbol{\beta}}_p, \widehat{\beta}_0) & \mathrm{V}(\widehat{\boldsymbol{\beta}}_p) \end{bmatrix}.$$

Therefore, the MSE matrix is

$$\mathrm{MSE}(\widehat{\boldsymbol{\beta}}) = \mathrm{V}(\widehat{\boldsymbol{\beta}}) = \sigma^2(\mathbf{X}'\mathbf{X})^{-1}$$
$$= \begin{bmatrix} \frac{\sigma^2}{n} & \mathbf{0} \\ \mathbf{0} & \sigma^2(\mathbf{X}'_p\mathbf{X}_p)^{-1} \end{bmatrix} = \begin{bmatrix} \mathrm{MSE}(\widehat{\beta}_0) & \mathbf{0} \\ \mathbf{0} & \mathrm{MSE}(\widehat{\boldsymbol{\beta}}_p) \end{bmatrix}, \quad (3.9)$$

and the TMSE is

$$\mathrm{TMSE}(\widehat{\boldsymbol{\beta}}) = trace\, \mathrm{MSE}(\widehat{\boldsymbol{\beta}}) = trace \begin{bmatrix} \mathrm{MSE}(\widehat{\beta}_0) & \mathbf{0} \\ \mathbf{0} & \mathrm{MSE}(\widehat{\boldsymbol{\beta}}_p) \end{bmatrix}$$
$$= \mathrm{MSE}(\widehat{\beta}_0) + \mathrm{TMSE}(\widehat{\boldsymbol{\beta}}_p)$$
$$= \frac{\sigma^2}{n} + \sigma^2 \sum_{i=1}^{n} \frac{1}{\lambda_i}. \quad (3.10)$$

Finally, we consider the MSE criteria of the OLS estimator

$$\widehat{\boldsymbol{\theta}} = \mathbf{T}^{-1}\widehat{\boldsymbol{\beta}} = \begin{bmatrix} 1 & -\mathbf{m}'_p\mathbf{S}_p^{-1} \\ \mathbf{0} & \mathbf{S}_p^{-1} \end{bmatrix} \begin{bmatrix} \widehat{\beta}_0 \\ \widehat{\boldsymbol{\beta}}_p \end{bmatrix} = \begin{bmatrix} \widehat{\beta}_0 - \mathbf{m}'_p\mathbf{S}_p^{-1}\widehat{\boldsymbol{\beta}}_p \\ \mathbf{S}_p^{-1}\widehat{\boldsymbol{\beta}}_p \end{bmatrix}.$$

The expectation vector and variance-covariance matrix are

$$\mathrm{E}(\widehat{\boldsymbol{\theta}}) = \mathrm{E}(\mathbf{T}^{-1}\widehat{\boldsymbol{\beta}}) = \mathbf{T}^{-1}\boldsymbol{\beta} = \boldsymbol{\theta} = \begin{bmatrix} \theta_0 \\ \boldsymbol{\theta}_p \end{bmatrix} \Leftrightarrow \mathrm{bias}(\widehat{\boldsymbol{\theta}}) = \mathbf{0} = \begin{bmatrix} 0 \\ \mathbf{0} \end{bmatrix} = \begin{bmatrix} \mathrm{bias}(\widehat{\theta}_0) \\ \mathrm{bias}(\widehat{\boldsymbol{\theta}}_p) \end{bmatrix},$$

18   3 MEAN SQUARED ERROR CRITERIA

$$\begin{aligned}
V(\widehat{\boldsymbol{\theta}}) &= V(\mathbf{T}^{-1}\widehat{\boldsymbol{\beta}}) = \mathbf{T}^{-1}V(\widehat{\boldsymbol{\beta}})(\mathbf{T}^{-1})' = \sigma^2 \mathbf{T}^{-1}(\mathbf{X}'\mathbf{X})^{-1}(\mathbf{T}^{-1})' \\
&= \begin{bmatrix} 1 & -\boldsymbol{m}'_p \mathbf{S}_p^{-1} \\ \mathbf{0} & \mathbf{S}_p^{-1} \end{bmatrix} \begin{bmatrix} \frac{\sigma^2}{n} & \mathbf{0} \\ \mathbf{0} & \sigma^2(\mathbf{X}'_p\mathbf{X}_p)^{-1} \end{bmatrix} \begin{bmatrix} 1 & \mathbf{0} \\ -\mathbf{S}_p^{-1}\boldsymbol{m}_p & \mathbf{S}_p^{-1} \end{bmatrix} \\
&= \begin{bmatrix} \frac{\sigma^2}{n} + \sigma^2 \boldsymbol{m}'_p \mathbf{S}_p^{-1}(\mathbf{X}'_p\mathbf{X}_p)^{-1}\mathbf{S}_p^{-1}\boldsymbol{m}_p & -\sigma^2 \boldsymbol{m}'_p \mathbf{S}_p^{-1}(\mathbf{X}'_p\mathbf{X}_p)^{-1} \\ -\sigma^2 \mathbf{S}_p^{-1}(\mathbf{X}'_p\mathbf{X}_p)^{-1}\mathbf{S}_p^{-1}\boldsymbol{m}_p & \sigma^2 \mathbf{S}_p^{-1}(\mathbf{X}'_p\mathbf{X}_p)^{-1}\mathbf{S}_p^{-1} \end{bmatrix} \\
&= \begin{bmatrix} V(\widehat{\theta}_0) & \mathrm{Cov}(\widehat{\theta}_0, \widehat{\boldsymbol{\theta}}_p) \\ \mathrm{Cov}(\widehat{\boldsymbol{\theta}}_p, \widehat{\theta}_0) & V(\widehat{\boldsymbol{\theta}}_p) \end{bmatrix}.
\end{aligned}$$

Therefore, the MSE matrix is

$$\begin{aligned}
\mathrm{MSE}(\widehat{\boldsymbol{\theta}}) &= V(\widehat{\boldsymbol{\theta}}) = \sigma^2 \mathbf{T}^{-1}(\mathbf{X}'\mathbf{X})^{-1}(\mathbf{T}^{-1})' \\
&= \begin{bmatrix} \frac{\sigma^2}{n} + \sigma^2 \boldsymbol{m}'_p \mathbf{S}_p^{-1}(\mathbf{X}'_p\mathbf{X}_p)^{-1}\mathbf{S}_p^{-1}\boldsymbol{m}_p & -\sigma^2 \boldsymbol{m}'_p \mathbf{S}_p^{-1}(\mathbf{X}'_p\mathbf{X}_p)^{-1} \\ -\sigma^2 \mathbf{S}_p^{-1}(\mathbf{X}'_p\mathbf{X}_p)^{-1}\mathbf{S}_p^{-1}\boldsymbol{m}_p & \sigma^2 \mathbf{S}_p^{-1}(\mathbf{X}'_p\mathbf{X}_p)^{-1}\mathbf{S}_p^{-1} \end{bmatrix} \\
&= \begin{bmatrix} \mathrm{MSE}(\widehat{\theta}_0) & \mathrm{MCE}(\widehat{\theta}_0, \widehat{\boldsymbol{\theta}}_p) \\ \mathrm{MCE}(\widehat{\boldsymbol{\theta}}_p, \widehat{\theta}_0) & \mathrm{MSE}(\widehat{\boldsymbol{\theta}}_p) \end{bmatrix},
\end{aligned} \quad (3.11)$$

and the TMSE is

$$\begin{aligned}
\mathrm{TMSE}(\widehat{\boldsymbol{\theta}}) &= trace\ \mathrm{MSE}(\widehat{\boldsymbol{\theta}}) = trace \begin{bmatrix} \mathrm{MSE}(\widehat{\theta}_0) & \mathrm{MCE}(\widehat{\theta}_0, \widehat{\boldsymbol{\theta}}_p) \\ \mathrm{MCE}(\widehat{\boldsymbol{\theta}}_p, \widehat{\theta}_0) & \mathrm{MSE}(\widehat{\boldsymbol{\theta}}_p) \end{bmatrix} \\
&= \mathrm{MSE}(\widehat{\theta}_0) + \mathrm{TMSE}(\widehat{\boldsymbol{\theta}}_p) \\
&= \left(\frac{\sigma^2}{n} + \sigma^2 \boldsymbol{m}'_p \mathbf{S}_p^{-1}(\mathbf{X}'_p\mathbf{X}_p)^{-1}\mathbf{S}_p^{-1}\boldsymbol{m}_p \right) + \sigma^2\ trace\ \mathbf{S}_p^{-1}(\mathbf{X}'_p\mathbf{X}_p)^{-1}\mathbf{S}_p^{-1}.
\end{aligned} \quad (3.12)$$

3.2.2 ORR Estimator

Next, we will discuss the MSE criteria of the ORR estimator. Specifically, we consider the following ORR estimator for $\boldsymbol{\beta}_p$:

$$\widehat{\boldsymbol{\beta}}_p(\cdot, k) = (\mathbf{X}'_p\mathbf{X}_p + k\mathbf{I}_p)^{-1}\mathbf{X}'_p\boldsymbol{Y}.$$

The expectation vector of the ORR estimator is

$$\begin{aligned}
\mathrm{E}(\widehat{\boldsymbol{\beta}}_p(\cdot, k)) &= (\mathbf{X}'_p\mathbf{X}_p + k\mathbf{I}_p)^{-1}\mathbf{X}'_p\mathrm{E}(\boldsymbol{Y}) = (\mathbf{X}'_p\mathbf{X}_p + k\mathbf{I}_p)^{-1}\mathbf{X}'_p\mathbf{X}\boldsymbol{\beta} \\
&= (\mathbf{X}'_p\mathbf{X}_p + k\mathbf{I}_p)^{-1}\mathbf{X}'_p\mathbf{X}_p\boldsymbol{\beta}_p = (\mathbf{I}_p + k(\mathbf{X}'_p\mathbf{X}_p)^{-1})^{-1}\boldsymbol{\beta}_p,
\end{aligned}$$

because of $\mathbf{X}'_p\mathbf{1} = \mathbf{0}$ and the bias vector of $\widehat{\boldsymbol{\beta}}_p(\cdot, k)$ is not the zero vector for $k > 0$:

$$\begin{aligned}
\mathrm{bias}(\widehat{\boldsymbol{\beta}}_p(\cdot, k)) &= \mathrm{E}(\widehat{\boldsymbol{\beta}}_p(\cdot, k)) - \boldsymbol{\beta}_p = (\mathbf{X}'_p\mathbf{X}_p + k\mathbf{I}_p)^{-1}\mathbf{X}'_p\mathbf{X}_p\boldsymbol{\beta}_p - \boldsymbol{\beta}_p \\
&= \{(\mathbf{X}'_p\mathbf{X}_p + k\mathbf{I}_p)^{-1}\mathbf{X}'_p\mathbf{X}_p - \mathbf{I}_p\}\boldsymbol{\beta}_p \neq \mathbf{0}.
\end{aligned}$$

3.2. MEAN SQUARED ERROR CRITERIA OF REGRESSION ESTIMATORS

The variance-covariance matrix of the ORR estimator is

$$\begin{aligned}V(\widehat{\boldsymbol{\beta}}_p(\cdot,k)) &= (\mathbf{X}_p'\mathbf{X}_p + k\mathbf{I}_p)^{-1}\mathbf{X}_p'V(\boldsymbol{Y})\mathbf{X}_p(\mathbf{X}_p'\mathbf{X}_p + k\mathbf{I}_p)^{-1} \\ &= \sigma^2(\mathbf{X}_p'\mathbf{X}_p + k\mathbf{I}_p)^{-1}\mathbf{X}_p'\mathbf{X}_p(\mathbf{X}_p'\mathbf{X}_p + k\mathbf{I}_p)^{-1}.\end{aligned}$$

Therefore, the MSE matrix is

$$\begin{aligned}\text{MSE}(\widehat{\boldsymbol{\beta}}_p(\cdot,k)) &= V(\widehat{\boldsymbol{\beta}}_p(\cdot,k)) + \text{bias}(\widehat{\boldsymbol{\beta}}_p(\cdot,k))\text{bias}(\widehat{\boldsymbol{\beta}}_p(\cdot,k))' \\ &= \sigma^2(\mathbf{X}_p'\mathbf{X}_p + k\mathbf{I}_p)^{-1}\mathbf{X}_p'\mathbf{X}_p(\mathbf{X}_p'\mathbf{X}_p + k\mathbf{I}_p)^{-1} \\ &\quad + \left\{(\mathbf{X}_p'\mathbf{X}_p + k\mathbf{I}_p)^{-1}\mathbf{X}_p'\mathbf{X}_p - \mathbf{I}_p\right\}\boldsymbol{\beta}_p\boldsymbol{\beta}_p'\left\{\mathbf{X}_p'\mathbf{X}_p(\mathbf{X}_p'\mathbf{X}_p + k\mathbf{I}_p)^{-1} - \mathbf{I}_p\right\},\end{aligned} \quad (3.13)$$

and the TMSE is

$$\begin{aligned}\text{TMSE}(\widehat{\boldsymbol{\beta}}_p(\cdot,k)) &= trace\,\text{MSE}(\widehat{\boldsymbol{\beta}}_p(\cdot,k)) \\ &= \sigma^2 trace(\mathbf{X}_p'\mathbf{X}_p + k\mathbf{I}_p)^{-1}\mathbf{X}_p'\mathbf{X}_p(\mathbf{X}_p'\mathbf{X}_p + k\mathbf{I}_p)^{-1} \\ &\quad + trace\left\{(\mathbf{X}_p'\mathbf{X}_p + k\mathbf{I}_p)^{-1}\mathbf{X}_p'\mathbf{X}_p - \mathbf{I}_p\right\}\boldsymbol{\beta}_p\boldsymbol{\beta}_p'\left\{\mathbf{X}_p'\mathbf{X}_p(\mathbf{X}_p'\mathbf{X}_p + k\mathbf{I}_p)^{-1} - \mathbf{I}_p\right\} \\ &= \sigma^2 trace\,\boldsymbol{\Gamma}_p\boldsymbol{\Lambda}_p(k)^{-1}\boldsymbol{\Lambda}_p\boldsymbol{\Lambda}_p(k)^{-1}\boldsymbol{\Gamma}_p' \\ &\quad + trace\,\boldsymbol{\Gamma}_p\left\{\boldsymbol{\Lambda}_p(k)^{-1}\boldsymbol{\Lambda}_p - \mathbf{I}_p\right\}\boldsymbol{\alpha}_p\boldsymbol{\alpha}_p'\left\{\boldsymbol{\Lambda}_p\boldsymbol{\Lambda}_p(k)^{-1} - \mathbf{I}_p\right\}\boldsymbol{\Gamma}_p' \\ &= \sigma^2 trace\,\boldsymbol{\Lambda}_p(k)^{-1}\boldsymbol{\Lambda}_p\boldsymbol{\Lambda}_p(k)^{-1} \\ &\quad + \boldsymbol{\alpha}_p'\left\{\boldsymbol{\Lambda}_p\boldsymbol{\Lambda}_p(k)^{-1} - \mathbf{I}_p\right\}\left\{\boldsymbol{\Lambda}_p(k)^{-1}\boldsymbol{\Lambda}_p - \mathbf{I}_p\right\}\boldsymbol{\alpha}_p \\ &= \sigma^2 \sum_{i=1}^p \frac{\lambda_i}{(\lambda_i + k)^2} + \sum_{i=1}^p \frac{k^2\alpha_i^2}{(\lambda_i + k)^2},\end{aligned} \quad (3.14)$$

where we use the following equalities:

$$\boldsymbol{\Lambda}_p(k) := \boldsymbol{\Lambda}_p + k\mathbf{I}_p,$$
$$(\mathbf{X}_p'\mathbf{X}_p + k\mathbf{I}_p)^{-1} = \boldsymbol{\Gamma}_p(\boldsymbol{\Lambda}_p + k\mathbf{I}_p)^{-1}\boldsymbol{\Gamma}_p' = \boldsymbol{\Gamma}_p\boldsymbol{\Lambda}_p(k)^{-1}\boldsymbol{\Gamma}_p',$$
$$\boldsymbol{\Gamma}_p'\boldsymbol{\Gamma}_p = \mathbf{I}_p, \quad \boldsymbol{\Gamma}_p'\boldsymbol{\beta}_p = \boldsymbol{\alpha}_p.$$

Next, we consider the MSE criteria of the ridge estimators

$$\begin{aligned}\widehat{\boldsymbol{\beta}}(\cdot, k_0) &= (\mathbf{X}'\mathbf{X} + \text{diag}(k_0, k, \cdots, k))^{-1}\mathbf{X}'\boldsymbol{Y} \\ &= \begin{bmatrix} n + k_0 & 0 \\ 0 & \mathbf{X}_p'\mathbf{X}_p + k\mathbf{I}_p \end{bmatrix}^{-1} \begin{bmatrix} \mathbf{1}' \\ \mathbf{X}_p' \end{bmatrix} \boldsymbol{Y} = \begin{bmatrix} \frac{n}{n+k_0}\overline{Y} \\ (\mathbf{X}_p'\mathbf{X}_p + k\mathbf{I}_p)^{-1}\mathbf{X}_p'\boldsymbol{Y} \end{bmatrix} \\ &= \begin{bmatrix} \widehat{\beta}_0(k_0) \\ \widehat{\boldsymbol{\beta}}_p(\cdot, k) \end{bmatrix}.\end{aligned}$$

Similar to the case $\widehat{\boldsymbol{\beta}}_p(\cdot, k)$,

$$\mathrm{E}(\widehat{\boldsymbol{\beta}}(\cdot, k_0)) = \begin{bmatrix} \mathrm{E}(\widehat{\beta}_0(k_0)) \\ \mathrm{E}(\widehat{\boldsymbol{\beta}}_p(\cdot, k)) \end{bmatrix} = \begin{bmatrix} \frac{n}{n+k_0}\beta_0 \\ (\mathbf{X}'_p\mathbf{X}_p + k\mathbf{I}_p)^{-1}\mathbf{X}'_p\mathbf{X}_p\boldsymbol{\beta}_p \end{bmatrix}$$

$$\Leftrightarrow \quad \mathrm{bias}(\widehat{\boldsymbol{\beta}}(\cdot, k_0)) = \begin{bmatrix} -\frac{k_0}{n+k_0}\beta_0 \\ \left\{(\mathbf{X}'_p\mathbf{X}_p + k\mathbf{I}_p)^{-1}\mathbf{X}'_p\mathbf{X}_p - \mathbf{I}_p\right\}\boldsymbol{\beta}_p \end{bmatrix} = \begin{bmatrix} \mathrm{bias}(\widehat{\beta}_0(k_0)) \\ \mathrm{bias}(\widehat{\boldsymbol{\beta}}_p(\cdot, k)) \end{bmatrix},$$

$$\mathrm{V}(\widehat{\boldsymbol{\beta}}(\cdot, k_0)) = \sigma^2 \left(\mathbf{X}'\mathbf{X} + \mathrm{diag}(k_0, k, \cdots, k)\right)^{-1} \mathbf{X}'\mathbf{X} \left(\mathbf{X}'\mathbf{X} + \mathrm{diag}(k_0, k, \cdots, k)\right)^{-1}$$

$$= \begin{bmatrix} \sigma^2 \frac{n}{(n+k_0)^2} & \mathbf{0} \\ \mathbf{0} & \sigma^2(\mathbf{X}'_p\mathbf{X}_p + k\mathbf{I}_p)^{-1}\mathbf{X}'_p\mathbf{X}_p(\mathbf{X}'_p\mathbf{X}_p + k\mathbf{I}_p)^{-1} \end{bmatrix}$$

$$= \begin{bmatrix} \mathrm{V}(\widehat{\beta}_0(k_0)) & \mathrm{Cov}(\widehat{\beta}_0(k_0), \widehat{\boldsymbol{\beta}}_p(\cdot, k)) \\ \mathrm{Cov}(\widehat{\boldsymbol{\beta}}_p(\cdot, k), \widehat{\beta}_0) & \mathrm{V}(\widehat{\boldsymbol{\beta}}_p(\cdot, k)) \end{bmatrix}.$$

Therefore, the MSE matrix is

$$\mathrm{MSE}(\widehat{\boldsymbol{\beta}}(\cdot, k_0)) = \mathrm{V}(\widehat{\boldsymbol{\beta}}(\cdot, k_0)) + \mathrm{bias}(\widehat{\boldsymbol{\beta}}(\cdot, k_0))\mathrm{bias}(\widehat{\boldsymbol{\beta}}(\cdot, k_0))'$$

$$= \begin{bmatrix} \sigma^2 \frac{n}{(n+k_0)^2} + \frac{k_0^2}{(n+k_0)^2}\beta_0^2 & -\frac{k_0}{n+k_0}\beta_0\boldsymbol{\beta}'_p\mathbf{X}'_p\mathbf{X}_p\left\{(\mathbf{X}'_p\mathbf{X}_p + k\mathbf{I}_p)^{-1} - \mathbf{I}_p\right\} \\ -\frac{k_0}{n+k_0}\beta_0\left\{(\mathbf{X}'_p\mathbf{X}_p + k\mathbf{I}_p)^{-1}\mathbf{X}'_p\mathbf{X}_p - \mathbf{I}_p\right\}\boldsymbol{\beta}_p & \mathrm{MSE}(\widehat{\boldsymbol{\beta}}_p(\cdot, k)) \end{bmatrix}$$

$$= \begin{bmatrix} \mathrm{MSE}(\widehat{\beta}_0(k_0)) & \mathrm{MCE}(\widehat{\beta}_0(k_0), \widehat{\boldsymbol{\beta}}_p(\cdot, k)) \\ \mathrm{MCE}(\widehat{\boldsymbol{\beta}}_p(\cdot, k), \widehat{\beta}_0(k_0)) & \mathrm{MSE}(\widehat{\boldsymbol{\beta}}_p(\cdot, k)) \end{bmatrix}, \quad (3.15)$$

and the TMSE is

$$\mathrm{TMSE}(\widehat{\boldsymbol{\beta}}(\cdot, k_0)) = trace\ \mathrm{MSE}(\widehat{\boldsymbol{\beta}}(\cdot, k_0))$$

$$= \mathrm{MSE}(\widehat{\beta}_0(k_0)) + \mathrm{TMSE}(\widehat{\boldsymbol{\beta}}_p(\cdot, k))$$

$$= \left(\sigma^2 \frac{n}{(n+k_0)^2} + \frac{k_0^2}{(n+k_0)^2}\beta_0^2\right) + \left(\sigma^2 \sum_{i=1}^p \frac{\lambda_i}{(\lambda_i+k)^2} + \sum_{i=1}^p \frac{k^2\alpha_i^2}{(\lambda_i+k)^2}\right). \quad (3.16)$$

Finally, we consider the MSE criteria of the ORR-type estimator

$$\widehat{\boldsymbol{\theta}}(\cdot, k_0) = \mathbf{T}^{-1}\widehat{\boldsymbol{\beta}}(\cdot, k_0) = \begin{bmatrix} 1 & -\boldsymbol{m}'_p\mathbf{S}_p^{-1} \\ \mathbf{0} & \mathbf{S}_p^{-1} \end{bmatrix} \begin{bmatrix} \widehat{\beta}_0(k_0) \\ \widehat{\boldsymbol{\beta}}_p(\cdot, k) \end{bmatrix}$$

$$= \begin{bmatrix} \widehat{\beta}_0(k_0) - \boldsymbol{m}'_p\mathbf{S}_p^{-1}\widehat{\boldsymbol{\beta}}_p(\cdot, k) \\ \mathbf{S}_p^{-1}\widehat{\boldsymbol{\beta}}_p(\cdot, k) \end{bmatrix} =: \begin{bmatrix} \widehat{\theta}_0(\cdot, k_0) \\ \widehat{\boldsymbol{\theta}}_p(\cdot, k) \end{bmatrix}.$$

3.2. MEAN SQUARED ERROR CRITERIA OF REGRESSION ESTIMATORS

The bias vector and variance-covariance matrix are

$$\text{bias}(\widehat{\boldsymbol{\theta}}(\cdot, k_0)) = \mathbf{T}^{-1}\text{bias}(\widehat{\boldsymbol{\beta}}(\cdot, k_0))$$
$$= \begin{bmatrix} 1 & -\boldsymbol{m}_p' \mathbf{S}_p^{-1} \\ \mathbf{0} & \mathbf{S}_p^{-1} \end{bmatrix} \begin{bmatrix} -\frac{k_0}{n+k_0}\beta_0 \\ \{(\mathbf{X}_p'\mathbf{X}_p + k\mathbf{I}_p)^{-1}\mathbf{X}_p'\mathbf{X}_p - \mathbf{I}_p\}\boldsymbol{\beta}_p \end{bmatrix}$$
$$= \begin{bmatrix} -\frac{k_0}{n+k_0}\beta_0 - \boldsymbol{m}_p' \mathbf{S}_p^{-1}\{(\mathbf{X}_p'\mathbf{X}_p + k\mathbf{I}_p)^{-1}\mathbf{X}_p'\mathbf{X}_p - \mathbf{I}_p\}\boldsymbol{\beta}_p \\ \mathbf{S}_p^{-1}\{(\mathbf{X}_p'\mathbf{X}_p + k\mathbf{I}_p)^{-1}\mathbf{X}_p'\mathbf{X}_p - \mathbf{I}_p\}\boldsymbol{\beta}_p \end{bmatrix}$$
$$=: \begin{bmatrix} \text{bias}(\widehat{\theta}_0(\cdot, k_0)) \\ \text{bias}(\widehat{\boldsymbol{\theta}}_p(\cdot, k)) \end{bmatrix},$$

and

$$V(\widehat{\boldsymbol{\theta}}(\cdot, k_0)) = \mathbf{T}^{-1} V(\widehat{\boldsymbol{\beta}}(\cdot, k_0))(\mathbf{T}^{-1})'$$
$$= \begin{bmatrix} V(\widehat{\theta}_0(\cdot, k_0)) & \text{Cov}(\widehat{\theta}_0(\cdot, k_0), \widehat{\boldsymbol{\theta}}_p(\cdot, k)) \\ \text{Cov}(\widehat{\boldsymbol{\theta}}_p(\cdot, k), \widehat{\theta}_0(\cdot, k_0)) & V(\widehat{\boldsymbol{\theta}}_p(\cdot, k)) \end{bmatrix},$$

where

$$V(\widehat{\theta}_0(\cdot, k_0)) = \sigma^2 \frac{n}{(n+k_0)^2}$$
$$+ \sigma^2 \boldsymbol{m}_p' \mathbf{S}_p^{-1}(\mathbf{X}_p'\mathbf{X}_p + k\mathbf{I}_p)^{-1}\mathbf{X}_p'\mathbf{X}_p(\mathbf{X}_p'\mathbf{X}_p + k\mathbf{I}_p)^{-1}\mathbf{S}_p^{-1}\boldsymbol{m}_p,$$
$$\text{Cov}(\widehat{\boldsymbol{\theta}}_p(\cdot, k), \widehat{\theta}_0(\cdot, k_0)) = \left\{\text{Cov}(\widehat{\theta}_0(\cdot, k_0), \widehat{\boldsymbol{\theta}}_p(\cdot, k))\right\}'$$
$$= -\sigma^2 \mathbf{S}_p^{-1}(\mathbf{X}_p'\mathbf{X}_p + k\mathbf{I}_p)^{-1}\mathbf{X}_p'\mathbf{X}_p(\mathbf{X}_p'\mathbf{X}_p + k\mathbf{I}_p)^{-1}\mathbf{S}_p^{-1}\boldsymbol{m}_p,$$
$$V(\widehat{\boldsymbol{\theta}}_p(\cdot, k)) = \sigma^2 \mathbf{S}_p^{-1}(\mathbf{X}_p'\mathbf{X}_p + k\mathbf{I}_p)^{-1}\mathbf{X}_p'\mathbf{X}_p(\mathbf{X}_p'\mathbf{X}_p + k\mathbf{I}_p)^{-1}\mathbf{S}_p^{-1}.$$

The MSE matrix is

$$\text{MSE}(\widehat{\boldsymbol{\theta}}(\cdot, k_0)) = V(\widehat{\boldsymbol{\theta}}(\cdot, k_0)) + \text{bias}(\widehat{\boldsymbol{\theta}}(\cdot, k_0))\text{bias}(\widehat{\boldsymbol{\theta}}(\cdot, k_0))'$$
$$= \begin{bmatrix} V(\widehat{\theta}_0(\cdot, k_0)) & \text{Cov}(\widehat{\theta}_0(\cdot, k_0), \widehat{\boldsymbol{\theta}}_p(\cdot, k)) \\ \text{Cov}(\widehat{\boldsymbol{\theta}}_p(\cdot, k), \widehat{\theta}_0(\cdot, k_0)) & V(\widehat{\boldsymbol{\theta}}_p(\cdot, k)) \end{bmatrix}$$
$$+ \begin{bmatrix} \text{bias}(\widehat{\theta}_0(\cdot, k_0))^2 & \text{bias}(\widehat{\theta}_0(\cdot, k_0))\text{bias}(\widehat{\boldsymbol{\theta}}_p(\cdot, k))' \\ \text{bias}(\widehat{\boldsymbol{\theta}}_p(\cdot, k))\text{bias}(\widehat{\theta}_0(\cdot, k_0)) & \text{bias}(\widehat{\boldsymbol{\theta}}_p(\cdot, k))\text{bias}(\widehat{\boldsymbol{\theta}}_p(\cdot, k))' \end{bmatrix}$$
$$= \begin{bmatrix} \text{MSE}(\widehat{\theta}_0(\cdot, k_0)) & \text{MCE}(\widehat{\theta}_0(\cdot, k_0), \widehat{\boldsymbol{\theta}}_p(\cdot, k)) \\ \text{MCE}(\widehat{\boldsymbol{\theta}}_p(\cdot, k), \widehat{\theta}_0(\cdot, k_0)) & \text{MSE}(\widehat{\boldsymbol{\theta}}_p(\cdot, k)) \end{bmatrix},$$

(3.17)

where

$$\begin{aligned}
\mathrm{MSE}(\widehat{\theta}_0(\cdot,k_0)) &= \mathrm{V}(\widehat{\theta}_0(\cdot,k_0)) + \mathrm{bias}(\widehat{\theta}_0(\cdot,k_0))^2 \\
&= \sigma^2 \frac{n}{(n+k_0)^2} \\
&\quad + \sigma^2 \boldsymbol{m}_p' \mathbf{S}_p^{-1} (\mathbf{X}_p'\mathbf{X}_p + k\mathbf{I}_p)^{-1} \mathbf{X}_p'\mathbf{X}_p (\mathbf{X}_p'\mathbf{X}_p + k\mathbf{I}_p)^{-1} \mathbf{S}_p^{-1} \boldsymbol{m}_p \\
&\quad + \left(-\frac{k_0}{n+k_0}\beta_0 - \boldsymbol{m}_p' \mathbf{S}_p^{-1} \left\{ (\mathbf{X}_p'\mathbf{X}_p + k\mathbf{I}_p)^{-1} \mathbf{X}_p'\mathbf{X}_p - \mathbf{I}_p \right\} \boldsymbol{\beta}_p \right)^2 \\
&= \sigma^2 \frac{n}{(n+k_0)^2} + \frac{k_0^2}{(n+k_0)^2}\beta_0^2 \\
&\quad + \sigma^2 \boldsymbol{m}_p' \mathbf{S}_p^{-1} (\mathbf{X}_p'\mathbf{X}_p + k\mathbf{I}_p)^{-1} \mathbf{X}_p'\mathbf{X}_p (\mathbf{X}_p'\mathbf{X}_p + k\mathbf{I}_p)^{-1} \mathbf{S}_p^{-1} \boldsymbol{m}_p \\
&\quad + 2\frac{k_0}{n+k_0}\beta_0 \boldsymbol{m}_p' \mathbf{S}_p^{-1} \left\{ (\mathbf{X}_p'\mathbf{X}_p + k\mathbf{I}_p)^{-1} \mathbf{X}_p'\mathbf{X}_p - \mathbf{I}_p \right\} \boldsymbol{\beta}_p \\
&\quad + \left(\boldsymbol{m}_p' \mathbf{S}_p^{-1} \left\{ (\mathbf{X}_p'\mathbf{X}_p + k\mathbf{I}_p)^{-1} \mathbf{X}_p'\mathbf{X}_p - \mathbf{I}_p \right\} \boldsymbol{\beta}_p \right)^2,
\end{aligned}$$

$$\begin{aligned}
\mathrm{MCE}(\widehat{\boldsymbol{\theta}}_p(\cdot,k), \widehat{\theta}_0(\cdot,k_0)) &= \mathrm{Cov}(\widehat{\boldsymbol{\theta}}_p(\cdot,k), \widehat{\theta}_0(\cdot,k_0)) + \mathrm{bias}(\widehat{\boldsymbol{\theta}}_p(\cdot,k))\mathrm{bias}(\widehat{\theta}_0(\cdot,k_0)) \\
&= -\sigma^2 \mathbf{S}_p^{-1} (\mathbf{X}_p'\mathbf{X}_p + k\mathbf{I}_p)^{-1} \mathbf{X}_p'\mathbf{X}_p (\mathbf{X}_p'\mathbf{X}_p + k\mathbf{I}_p)^{-1} \mathbf{S}_p^{-1} \boldsymbol{m}_p \\
&\quad + \mathbf{S}_p^{-1} \left\{ (\mathbf{X}_p'\mathbf{X}_p + k\mathbf{I}_p)^{-1} \mathbf{X}_p'\mathbf{X}_p - \mathbf{I}_p \right\} \boldsymbol{\beta}_p \\
&\quad \times \left\{ -\frac{k_0}{n+k_0}\beta_0 - \boldsymbol{m}_p' \mathbf{S}_p^{-1} \left\{ (\mathbf{X}_p'\mathbf{X}_p + k\mathbf{I}_p)^{-1} \mathbf{X}_p'\mathbf{X}_p - \mathbf{I}_p \right\} \boldsymbol{\beta}_p \right.
\end{aligned}$$

$$\begin{aligned}
\mathrm{MSE}(\widehat{\boldsymbol{\theta}}_p(\cdot,k)) &= \mathrm{V}(\widehat{\boldsymbol{\theta}}_p(\cdot,k)) + \mathrm{bias}(\widehat{\boldsymbol{\theta}}_p(\cdot,k))\mathrm{bias}(\widehat{\boldsymbol{\theta}}_p(\cdot,k))' \\
&= \sigma^2 \mathbf{S}_p^{-1} (\mathbf{X}_p'\mathbf{X}_p + k\mathbf{I}_p)^{-1} \mathbf{X}_p'\mathbf{X}_p (\mathbf{X}_p'\mathbf{X}_p + k\mathbf{I}_p)^{-1} \mathbf{S}_p^{-1} \\
&\quad + \mathbf{S}_p^{-1} \left\{ (\mathbf{X}_p'\mathbf{X}_p + k\mathbf{I}_p)^{-1} \mathbf{X}_p'\mathbf{X}_p - \mathbf{I}_p \right\} \boldsymbol{\beta}_p \\
&\quad \times \left\{ \mathbf{S}_p^{-1} \left\{ (\mathbf{X}_p'\mathbf{X}_p + k\mathbf{I}_p)^{-1} \mathbf{X}_p'\mathbf{X}_p - \mathbf{I}_p \right\} \boldsymbol{\beta}_p \right\}' \\
&= \sigma^2 \mathbf{S}_p^{-1} (\mathbf{X}_p'\mathbf{X}_p + k\mathbf{I}_p)^{-1} \mathbf{X}_p'\mathbf{X}_p (\mathbf{X}_p'\mathbf{X}_p + k\mathbf{I}_p)^{-1} \mathbf{S}_p^{-1} \\
&\quad + \mathbf{S}_p^{-1} \left\{ (\mathbf{X}_p'\mathbf{X}_p + k\mathbf{I}_p)^{-1} \mathbf{X}_p'\mathbf{X}_p - \mathbf{I}_p \right\} \boldsymbol{\beta}_p \\
&\quad \times \boldsymbol{\beta}_p' \left\{ \mathbf{X}_p'\mathbf{X}_p (\mathbf{X}_p'\mathbf{X}_p + k\mathbf{I}_p)^{-1} - \mathbf{I}_p \right\} \mathbf{S}_p^{-1}.
\end{aligned}$$

Consequently, the TMSE is

$$\begin{aligned}
&\text{TMSE}(\widehat{\boldsymbol{\theta}}(\cdot, k_0)) \\
&= \text{trace MSE}(\widehat{\boldsymbol{\theta}}(\cdot, k_0)) = \text{MSE}(\widehat{\theta}_0(\cdot, k_0)) + \text{trace MSE}(\widehat{\boldsymbol{\theta}}_p(\cdot, k)) \\
&= \Big\{ \sigma^2 \frac{n}{(n+k_0)^2} + \sigma^2 \boldsymbol{m}_p' \mathbf{S}_p^{-1} (\mathbf{X}_p' \mathbf{X}_p + k\mathbf{I}_p)^{-1} \mathbf{X}_p' \mathbf{X}_p (\mathbf{X}_p' \mathbf{X}_p + k\mathbf{I}_p)^{-1} \mathbf{S}_p^{-1} \boldsymbol{m}_p \\
&\quad + \frac{k_0^2}{(n+k_0)^2} \beta_0^2 + 2 \frac{k_0}{n+k_0} \beta_0 \boldsymbol{m}_p' \mathbf{S}_p^{-1} \left\{ (\mathbf{X}_p' \mathbf{X}_p + k\mathbf{I}_p)^{-1} \mathbf{X}_p' \mathbf{X}_p - \mathbf{I}_p \right\} \boldsymbol{\beta}_p \\
&\quad + \left(\boldsymbol{m}_p' \mathbf{S}_p^{-1} \left\{ (\mathbf{X}_p' \mathbf{X}_p + k\mathbf{I}_p)^{-1} \mathbf{X}_p' \mathbf{X}_p - \mathbf{I}_p \right\} \boldsymbol{\beta}_p \right)^2 \Big\} \\
&\quad + \sigma^2 \text{trace } \mathbf{S}_p^{-1} (\mathbf{X}_p' \mathbf{X}_p + k\mathbf{I}_p)^{-1} \mathbf{X}_p' \mathbf{X}_p (\mathbf{X}_p' \mathbf{X}_p + k\mathbf{I}_p)^{-1} \mathbf{S}_p^{-1} \\
&\quad + \text{trace } \mathbf{S}_p^{-1} \left\{ (\mathbf{X}_p' \mathbf{X}_p + k\mathbf{I}_p)^{-1} \mathbf{X}_p' \mathbf{X}_p - \mathbf{I}_p \right\} \boldsymbol{\beta}_p \boldsymbol{\beta}_p' \left\{ \mathbf{X}_p' \mathbf{X}_p (\mathbf{X}_p' \mathbf{X}_p + k\mathbf{I}_p)^{-1} - \mathbf{I}_p \right\} \mathbf{S}_p^{-1} \\
&= \sigma^2 \frac{n}{(n+k_0)^2} + \frac{k_0^2}{(n+k_0)^2} \beta_0^2 \\
&\quad + 2 \frac{k_0}{n+k_0} \beta_0 \boldsymbol{m}_p' \mathbf{S}_p^{-1} \left\{ (\mathbf{X}_p' \mathbf{X}_p + k\mathbf{I}_p)^{-1} \mathbf{X}_p' \mathbf{X}_p - \mathbf{I}_p \right\} \boldsymbol{\beta}_p \\
&\quad + \sigma^2 \text{trace } (\mathbf{X}_p' \mathbf{X}_p + k\mathbf{I}_p)^{-1} \mathbf{X}_p' \mathbf{X}_p (\mathbf{X}_p' \mathbf{X}_p + k\mathbf{I}_p)^{-1} \left(\mathbf{S}_p^{-1} \boldsymbol{m}_p \boldsymbol{m}_p' \mathbf{S}_p^{-1} + \mathbf{S}_p^{-2} \right) \\
&\quad + \boldsymbol{\beta}_p' \left\{ \mathbf{X}_p' \mathbf{X}_p (\mathbf{X}_p' \mathbf{X}_p + k\mathbf{I}_p)^{-1} - \mathbf{I}_p \right\} \left(\mathbf{S}_p^{-1} \boldsymbol{m}_p \boldsymbol{m}_p' \mathbf{S}_p^{-1} + \mathbf{S}_p^{-2} \right) \left\{ (\mathbf{X}_p' \mathbf{X}_p + k\mathbf{I}_p)^{-1} \mathbf{X}_p' \mathbf{X}_p - \mathbf{I}_p \right\} \boldsymbol{\beta}_p \\
&= \sigma^2 \frac{n}{(n+k_0)^2} + \frac{k_0^2}{(n+k_0)^2} \alpha_0^2 \\
&\quad + 2 \frac{k_0}{n+k_0} \alpha_0 \boldsymbol{m}_p' \mathbf{S}_p^{-1} \boldsymbol{\Gamma}_p \left\{ \boldsymbol{\Lambda}_p(k)^{-1} \boldsymbol{\Lambda}_p - \mathbf{I}_p \right\} \boldsymbol{\alpha}_p \\
&\quad + \sigma^2 \text{trace } \boldsymbol{\Lambda}_p(k)^{-1} \boldsymbol{\Lambda}_p \boldsymbol{\Lambda}_p(k)^{-1} \boldsymbol{\Gamma}_p' \left(\mathbf{S}_p^{-1} \boldsymbol{m}_p \boldsymbol{m}_p' \mathbf{S}_p^{-1} + \mathbf{S}_p^{-2} \right) \boldsymbol{\Gamma}_p \\
&\quad + \boldsymbol{\alpha}_p' \left\{ \boldsymbol{\Lambda}_p \boldsymbol{\Lambda}_p(k)^{-1} - \mathbf{I}_p \right\} \boldsymbol{\Gamma}_p' \left(\mathbf{S}_p^{-1} \boldsymbol{m}_p \boldsymbol{m}_p' \mathbf{S}_p^{-1} + \mathbf{S}_p^{-2} \right) \boldsymbol{\Gamma}_p \left\{ \boldsymbol{\Lambda}_p(k)^{-1} \boldsymbol{\Lambda}_p - \mathbf{I}_p \right\} \boldsymbol{\alpha}_p,
\end{aligned} \tag{3.18}$$

where

$$\boldsymbol{\alpha} = \boldsymbol{\Gamma}' \boldsymbol{\beta} = \begin{bmatrix} 1 & \mathbf{0} \\ \mathbf{0} & \boldsymbol{\Gamma}_p' \end{bmatrix} \begin{bmatrix} \beta_0 \\ \boldsymbol{\beta}_p \end{bmatrix} = \begin{bmatrix} \beta_0 \\ \boldsymbol{\Gamma}_p' \boldsymbol{\beta}_p \end{bmatrix} = \begin{bmatrix} \alpha_0 \\ \boldsymbol{\alpha}_p \end{bmatrix}.$$

3.2.3 GRR Estimator

We turn here to the MSE criteria of the GRR estimator. We start by considering the following GRR estimator for $\boldsymbol{\beta}_p$:

$$\widehat{\boldsymbol{\beta}}_p(\mathbf{K}_p) = (\mathbf{X}_p' \mathbf{X}_p + \boldsymbol{\Gamma}_p \mathbf{K}_p \boldsymbol{\Gamma}_p')^{-1} \mathbf{X}_p' \mathbf{Y}.$$

The expectation of the GRR estimator is

$$\begin{aligned}
\text{E}(\widehat{\boldsymbol{\beta}}_p(\mathbf{K}_p)) &= (\mathbf{X}_p' \mathbf{X}_p + \boldsymbol{\Gamma}_p \mathbf{K}_p \boldsymbol{\Gamma}_p')^{-1} \mathbf{X}_p' \text{E}(\mathbf{Y}) = (\mathbf{X}_p' \mathbf{X}_p + \boldsymbol{\Gamma}_p \mathbf{K}_p \boldsymbol{\Gamma}_p')^{-1} \mathbf{X}_p' \mathbf{X} \boldsymbol{\beta} \\
&= (\mathbf{X}_p' \mathbf{X}_p + \boldsymbol{\Gamma}_p \mathbf{K}_p \boldsymbol{\Gamma}_p')^{-1} \mathbf{X}_p' \mathbf{X}_p \boldsymbol{\beta}_p,
\end{aligned}$$

because $\mathbf{X}_p' \mathbf{1} = \mathbf{0}$ and the bias vector of $\widehat{\boldsymbol{\beta}}_p(\mathbf{K}_p)$ is not the zero vector for

$k > 0$,

$$\text{bias}(\widehat{\boldsymbol{\beta}}_p(\mathbf{K}_p)) = \text{E}(\widehat{\boldsymbol{\beta}}_p(\mathbf{K}_p)) - \boldsymbol{\beta}_p = (\mathbf{X}_p'\mathbf{X}_p + \boldsymbol{\Gamma}_p\mathbf{K}_p\boldsymbol{\Gamma}_p')^{-1}\mathbf{X}_p'\mathbf{X}_p\boldsymbol{\beta}_p - \boldsymbol{\beta}_p$$
$$= \left\{(\mathbf{X}_p'\mathbf{X}_p + \boldsymbol{\Gamma}_p\mathbf{K}_p\boldsymbol{\Gamma}_p')^{-1}\mathbf{X}_p'\mathbf{X}_p - \mathbf{I}_p\right\}\boldsymbol{\beta}_p \neq \mathbf{0}.$$

The variance-covariance matrix of the OLS estimator is

$$\text{V}(\widehat{\boldsymbol{\beta}}_p(\mathbf{K}_p)) = (\mathbf{X}_p'\mathbf{X}_p + \boldsymbol{\Gamma}_p\mathbf{K}_p\boldsymbol{\Gamma}_p')^{-1}\mathbf{X}_p'\text{V}(Y)\mathbf{X}_p(\mathbf{X}_p'\mathbf{X}_p + \boldsymbol{\Gamma}_p\mathbf{K}_p\boldsymbol{\Gamma}_p')^{-1}$$
$$= \sigma^2(\mathbf{X}_p'\mathbf{X}_p + \boldsymbol{\Gamma}_p\mathbf{K}_p\boldsymbol{\Gamma}_p')^{-1}\mathbf{X}_p'\mathbf{X}_p(\mathbf{X}_p'\mathbf{X}_p + \boldsymbol{\Gamma}_p\mathbf{K}_p\boldsymbol{\Gamma}_p')^{-1}.$$

Therefore, the MSE matrix is

$$\text{MSE}(\widehat{\boldsymbol{\beta}}_p(\mathbf{K}_p)) = \text{V}(\widehat{\boldsymbol{\beta}}_p(\mathbf{K}_p)) + \text{bias}(\widehat{\boldsymbol{\beta}}_p(\mathbf{K}_p))\text{bias}(\widehat{\boldsymbol{\beta}}_p(\mathbf{K}_p))'$$
$$= \sigma^2(\mathbf{X}_p'\mathbf{X}_p + \boldsymbol{\Gamma}_p\mathbf{K}_p\boldsymbol{\Gamma}_p')^{-1}\mathbf{X}_p'\mathbf{X}_p(\mathbf{X}_p'\mathbf{X}_p + \boldsymbol{\Gamma}_p\mathbf{K}_p\boldsymbol{\Gamma}_p')^{-1}$$
$$+ \left\{(\mathbf{X}_p'\mathbf{X}_p + \boldsymbol{\Gamma}_p\mathbf{K}_p\boldsymbol{\Gamma}_p')^{-1}\mathbf{X}_p'\mathbf{X}_p - \mathbf{I}_p\right\}\boldsymbol{\beta}_p$$
$$\times \boldsymbol{\beta}_p'\left\{\mathbf{X}_p'\mathbf{X}_p(\mathbf{X}_p'\mathbf{X}_p + \boldsymbol{\Gamma}_p\mathbf{K}_p\boldsymbol{\Gamma}_p')^{-1} - \mathbf{I}_p\right\}, \quad (3.19)$$

and the TMSE is

$$\text{TMSE}(\widehat{\boldsymbol{\beta}}_p(\mathbf{K}_p)) = \textit{trace}\ \text{MSE}(\widehat{\boldsymbol{\beta}}_p(\mathbf{K}_p))$$
$$= \sigma^2 \textit{trace}(\mathbf{X}_p'\mathbf{X}_p + \boldsymbol{\Gamma}_p\mathbf{K}_p\boldsymbol{\Gamma}_p')^{-1}\mathbf{X}_p'\mathbf{X}_p(\mathbf{X}_p'\mathbf{X}_p + \boldsymbol{\Gamma}_p\mathbf{K}_p\boldsymbol{\Gamma}_p')^{-1}$$
$$+ \textit{trace}\left(\left\{(\mathbf{X}_p'\mathbf{X}_p + \boldsymbol{\Gamma}_p\mathbf{K}_p\boldsymbol{\Gamma}_p')^{-1}\mathbf{X}_p'\mathbf{X}_p - \mathbf{I}_p\right\}\boldsymbol{\beta}_p\right.$$
$$\left.\times \boldsymbol{\beta}_p'\left\{\mathbf{X}_p'\mathbf{X}_p(\mathbf{X}_p'\mathbf{X}_p + \boldsymbol{\Gamma}_p\mathbf{K}_p\boldsymbol{\Gamma}_p')^{-1} - \mathbf{I}_p\right\}\right)$$
$$= \sigma^2 \textit{trace}\boldsymbol{\Gamma}_p\boldsymbol{\Lambda}_p(\mathbf{K}_p)^{-1}\boldsymbol{\Lambda}_p\boldsymbol{\Lambda}_p(\mathbf{K}_p)^{-1}\boldsymbol{\Gamma}_p'$$
$$+ \textit{trace}\boldsymbol{\Gamma}_p\left\{\boldsymbol{\Lambda}_p(\mathbf{K}_p)^{-1}\boldsymbol{\Lambda}_p - \mathbf{I}_p\right\}\boldsymbol{\alpha}_p\boldsymbol{\alpha}_p'\left\{\boldsymbol{\Lambda}_p\boldsymbol{\Lambda}_p(\mathbf{K}_p)^{-1} - \mathbf{I}_p\right\}\boldsymbol{\Gamma}_p'$$
$$= \sigma^2 \textit{trace}\boldsymbol{\Lambda}_p(\mathbf{K}_p)^{-1}\boldsymbol{\Lambda}_p\boldsymbol{\Lambda}_p(\mathbf{K}_p)^{-1}$$
$$+ \boldsymbol{\alpha}_p'\left\{\boldsymbol{\Lambda}_p\boldsymbol{\Lambda}_p(\mathbf{K}_p)^{-1} - \mathbf{I}_p\right\}\left\{\boldsymbol{\Lambda}_p(\mathbf{K}_p)^{-1}\boldsymbol{\Lambda}_p - \mathbf{I}_p\right\}\boldsymbol{\alpha}_p$$
$$= \sigma^2 \sum_{i=1}^{p}\frac{\lambda_i}{(\lambda_i + k_i)^2} + \sum_{i=1}^{p}\frac{k_i^2\alpha_i^2}{(\lambda_i + k_i)^2}, \quad (3.20)$$

where we use the following equalities:

$$\boldsymbol{\Lambda}_p(\mathbf{K}_p) := \boldsymbol{\Lambda}_p + \mathbf{K}_p, \quad (\mathbf{X}_p'\mathbf{X}_p + \boldsymbol{\Gamma}_p\mathbf{K}_p\boldsymbol{\Gamma}_p')^{-1} = \boldsymbol{\Gamma}_p\boldsymbol{\Lambda}_p(\mathbf{K}_p)^{-1}\boldsymbol{\Gamma}_p'.$$

Next, we consider the MSE criteria of the GRR estimators

$$\widehat{\boldsymbol{\beta}}(\mathbf{K}) = (\mathbf{X}'\mathbf{X} + \boldsymbol{\Gamma}\mathbf{K}\boldsymbol{\Gamma})^{-1}\mathbf{X}'Y$$
$$= \begin{bmatrix} n + k_0 & 0 \\ 0 & \mathbf{X}_p'\mathbf{X}_p + \boldsymbol{\Gamma}_p\mathbf{K}_p\boldsymbol{\Gamma}_p' \end{bmatrix}^{-1}\begin{bmatrix} \mathbf{1}' \\ \mathbf{X}_p' \end{bmatrix}Y$$
$$= \begin{bmatrix} \frac{n}{n+k_0}\overline{Y} \\ (\mathbf{X}_p'\mathbf{X}_p + \boldsymbol{\Gamma}_p\mathbf{K}_p\boldsymbol{\Gamma}_p')^{-1}\mathbf{X}_p'Y \end{bmatrix} = \begin{bmatrix} \widehat{\beta}_0(k_0) \\ \widehat{\boldsymbol{\beta}}_p(\mathbf{K}_p) \end{bmatrix},$$

3.2. MEAN SQUARED ERROR CRITERIA OF REGRESSION ESTIMATORS

where $\mathbf{K} := \text{diag}(k_0, k_1, \cdots, k_p)$.

Similar to the case $\widehat{\boldsymbol{\beta}}_p(\mathbf{K}_p)$,

$$\text{E}(\widehat{\boldsymbol{\beta}}(\mathbf{K})) = \begin{bmatrix} \text{E}(\widehat{\beta}_0(k_0)) \\ \text{E}(\widehat{\boldsymbol{\beta}}_p(\mathbf{K}_p)) \end{bmatrix} = \begin{bmatrix} \frac{n}{n+k_0}\beta_0 \\ (\mathbf{X}'_p\mathbf{X}_p + \boldsymbol{\Gamma}_p\mathbf{K}_p\boldsymbol{\Gamma}'_p)^{-1}\mathbf{X}'_p\mathbf{X}_p\boldsymbol{\beta}_p \end{bmatrix},$$

$$\text{bias}(\widehat{\boldsymbol{\beta}}(\mathbf{K})) = \begin{bmatrix} -\frac{k_0}{n+k_0}\beta_0 \\ \{(\mathbf{X}'_p\mathbf{X}_p + \boldsymbol{\Gamma}_p\mathbf{K}_p\boldsymbol{\Gamma}'_p)^{-1}\mathbf{X}'_p\mathbf{X}_p - \mathbf{I}_p\}\boldsymbol{\beta}_p \end{bmatrix}$$

$$= \begin{bmatrix} \text{bias}(\widehat{\beta}_0(k_0)) \\ \text{bias}(\widehat{\boldsymbol{\beta}}_p(\mathbf{K}_p)) \end{bmatrix},$$

$$\text{V}(\widehat{\boldsymbol{\beta}}(\mathbf{K})) = \sigma^2 (\mathbf{X}'\mathbf{X} + \boldsymbol{\Gamma}\mathbf{K}\boldsymbol{\Gamma}')^{-1} \mathbf{X}'\mathbf{X} (\mathbf{X}'\mathbf{X} + \boldsymbol{\Gamma}\mathbf{K}\boldsymbol{\Gamma}')^{-1}$$

$$= \begin{bmatrix} \sigma^2 \frac{n}{(n+k_0)^2} & 0 \\ 0 & \sigma^2(\mathbf{X}'_p\mathbf{X}_p + \boldsymbol{\Gamma}_p\mathbf{K}_p\boldsymbol{\Gamma}'_p)^{-1}\mathbf{X}'_p\mathbf{X}_p(\mathbf{X}'_p\mathbf{X}_p + \boldsymbol{\Gamma}_p\mathbf{K}_p\boldsymbol{\Gamma}'_p)^{-1} \end{bmatrix}$$

$$= \begin{bmatrix} \text{V}(\widehat{\beta}_0(k_0)) & \text{Cov}(\widehat{\beta}_0(k_0), \widehat{\boldsymbol{\beta}}_p(\mathbf{K}_p)) \\ \text{Cov}(\widehat{\boldsymbol{\beta}}_p(\mathbf{K}_p), \widehat{\beta}_0(k_0)) & \text{V}(\widehat{\boldsymbol{\beta}}_p(\mathbf{K}_p)) \end{bmatrix}.$$

Therefore, the MSE matrix is

$$\text{MSE}(\widehat{\boldsymbol{\beta}}(\mathbf{K})) = \text{V}(\widehat{\boldsymbol{\beta}}(\mathbf{K})) + \text{bias}(\widehat{\boldsymbol{\beta}}(\mathbf{K}))\text{bias}(\widehat{\boldsymbol{\beta}}(\mathbf{K}))'$$

$$= \begin{bmatrix} \text{V}(\widehat{\beta}_0(k_0)) & \text{Cov}(\widehat{\beta}_0(k_0), \widehat{\boldsymbol{\beta}}_p(\mathbf{K}_p)) \\ \text{Cov}(\widehat{\boldsymbol{\beta}}_p(\mathbf{K}_p), \widehat{\beta}_0(k_0)) & \text{V}(\widehat{\boldsymbol{\beta}}_p(\mathbf{K}_p)) \end{bmatrix} \quad (3.21)$$

$$+ \begin{bmatrix} \text{bias}(\widehat{\beta}_0(k_0))^2 & \text{bias}(\widehat{\beta}_0(k_0))\text{bias}(\widehat{\boldsymbol{\beta}}_p(\mathbf{K}_p))' \\ \text{bias}(\widehat{\boldsymbol{\beta}}_p(\mathbf{K}_p))\text{bias}(\widehat{\beta}_0(k_0)) & \text{bias}(\widehat{\boldsymbol{\beta}}_p(\mathbf{K}_p))\text{bias}(\widehat{\boldsymbol{\beta}}_p(\mathbf{K}_p))' \end{bmatrix}$$

$$= \begin{bmatrix} \text{MSE}(\widehat{\beta}_0(k_0)) & \text{MCE}(\widehat{\beta}_0(k_0), \widehat{\boldsymbol{\beta}}_p(\mathbf{K}_p)) \\ \text{MCE}(\widehat{\boldsymbol{\beta}}_p(\mathbf{K}_p), \widehat{\beta}_0(k_0)) & \text{MSE}(\widehat{\boldsymbol{\beta}}_p(\mathbf{K}_p)) \end{bmatrix}, \quad (3.22)$$

where

$$\text{MSE}(\widehat{\beta}_0(k_0)) = \text{V}(\widehat{\beta}_0(k_0)) + \text{bias}(\widehat{\beta}_0(k_0))^2$$

$$= \sigma^2 \frac{n}{(n+k_0)^2} + \frac{k_0^2}{(n+k_0)^2}\beta_0^2,$$

$$\text{MCE}(\widehat{\boldsymbol{\beta}}_p(\mathbf{K}_p), \widehat{\beta}_0(k_0)) = \text{Cov}(\widehat{\boldsymbol{\beta}}_p(\mathbf{K}_p), \widehat{\beta}_0(k_0)) + \text{bias}(\widehat{\boldsymbol{\beta}}_p(\mathbf{K}_p))\text{bias}(\widehat{\beta}_0(k_0))$$

$$= -\frac{k_0}{n+k_0}\beta_0\{(\mathbf{X}'_p\mathbf{X}_p + \boldsymbol{\Gamma}_p\mathbf{K}_p\boldsymbol{\Gamma}'_p)^{-1}\mathbf{X}'_p\mathbf{X}_p - \mathbf{I}_p\}\boldsymbol{\beta}_p$$

$$= \text{MCE}(\widehat{\beta}_0(k_0), \widehat{\boldsymbol{\beta}}_p(\mathbf{K}_p))',$$

and the TMSE is

$$\begin{aligned}\text{TMSE}(\widehat{\boldsymbol{\beta}}(\mathbf{K})) &= \textit{trace } \text{MSE}(\widehat{\boldsymbol{\beta}}(\mathbf{K})) \\ &= \text{MSE}(\widehat{\beta}_0(k_0)) + \text{TMSE}(\widehat{\boldsymbol{\beta}}_p(\mathbf{K}_p)) \\ &= \left(\sigma^2 \frac{n}{(n+k_0)^2} + \frac{k_0^2 \alpha_0^2}{(n+k_0)^2}\right) + \left(\sigma^2 \sum_{i=1}^p \frac{\lambda_i}{(\lambda_i+k_i)^2} + \sum_{i=1}^p \frac{k_i^2 \alpha_i^2}{(\lambda_i+k_i)^2}\right). \end{aligned}$$
(3.23)

Finally, we consider the MSE criteria of the GRR-type estimator

$$\begin{aligned}\widehat{\boldsymbol{\theta}}(\mathbf{K}) = \mathbf{T}^{-1}\widehat{\boldsymbol{\beta}}(\mathbf{K}) &= \begin{bmatrix} 1 & -\boldsymbol{m}_p' \mathbf{S}_p^{-1} \\ 0 & \mathbf{S}_p^{-1} \end{bmatrix} \begin{bmatrix} \widehat{\beta}_0(k_0) \\ \widehat{\boldsymbol{\beta}}_p(\mathbf{K}_p) \end{bmatrix} \\ &= \begin{bmatrix} \widehat{\beta}_0(k_0) - \boldsymbol{m}_p' \mathbf{S}_p^{-1} \widehat{\boldsymbol{\beta}}_p(\mathbf{K}_p) \\ \mathbf{S}_p^{-1} \widehat{\boldsymbol{\beta}}_p(\mathbf{K}_p) \end{bmatrix} =: \begin{bmatrix} \widehat{\theta}_0(\mathbf{K}) \\ \widehat{\boldsymbol{\theta}}_p(\mathbf{K}_p) \end{bmatrix}. \end{aligned}$$

The bias vector and variance-covariance matrix are

$$\begin{aligned}\text{bias}(\widehat{\boldsymbol{\theta}}(\mathbf{K})) &= \mathbf{T}^{-1} \text{bias}(\widehat{\boldsymbol{\beta}}(\mathbf{K})) \\ &= \begin{bmatrix} 1 & -\boldsymbol{m}_p' \mathbf{S}_p^{-1} \\ 0 & \mathbf{S}_p^{-1} \end{bmatrix} \begin{bmatrix} -\frac{k_0}{n+k_0}\beta_0 \\ \{(\mathbf{X}_p'\mathbf{X}_p + \boldsymbol{\Gamma}_p \mathbf{K}_p \boldsymbol{\Gamma}_p')^{-1}\mathbf{X}_p'\mathbf{X}_p - \mathbf{I}_p\}\boldsymbol{\beta}_p \end{bmatrix} \\ &= \begin{bmatrix} -\frac{k_0}{n+k_0}\beta_0 - \boldsymbol{m}_p'\mathbf{S}_p^{-1}\{(\mathbf{X}_p'\mathbf{X}_p + \boldsymbol{\Gamma}_p \mathbf{K}_p \boldsymbol{\Gamma}_p')^{-1}\mathbf{X}_p'\mathbf{X}_p - \mathbf{I}_p\}\boldsymbol{\beta}_p \\ \mathbf{S}_p^{-1}\{(\mathbf{X}_p'\mathbf{X}_p + \boldsymbol{\Gamma}_p \mathbf{K}_p \boldsymbol{\Gamma}_p')^{-1}\mathbf{X}_p'\mathbf{X}_p - \mathbf{I}_p\}\boldsymbol{\beta}_p \end{bmatrix} \\ &=: \begin{bmatrix} \text{bias}(\widehat{\theta}_0(\mathbf{K})) \\ \text{bias}(\widehat{\boldsymbol{\theta}}_p(\mathbf{K}_p)) \end{bmatrix}, \end{aligned}$$

and

$$\mathrm{V}(\widehat{\boldsymbol{\theta}}(\mathbf{K})) = \mathbf{T}^{-1}\mathrm{V}(\widehat{\boldsymbol{\beta}}(\mathbf{K}))(\mathbf{T}^{-1})' = \begin{bmatrix} \mathrm{V}(\widehat{\theta}_0(\mathbf{K})) & \mathrm{Cov}(\widehat{\theta}_0(\mathbf{K}), \widehat{\boldsymbol{\theta}}_p(\mathbf{K}_p)) \\ \mathrm{Cov}(\widehat{\boldsymbol{\theta}}_p(\mathbf{K}_p), \widehat{\theta}_0(\mathbf{K})) & \mathrm{V}(\widehat{\boldsymbol{\theta}}_p(\mathbf{K}_p)) \end{bmatrix},$$

where

$$\mathrm{V}(\widehat{\theta}_0(\mathbf{K}))$$
$$= \sigma^2 \frac{n}{(n+k_0)^2} + \sigma^2 \boldsymbol{m}_p' \mathbf{S}_p^{-1}(\mathbf{X}_p'\mathbf{X}_p + \boldsymbol{\Gamma}_p \mathbf{K}_p \boldsymbol{\Gamma}_p')^{-1}\mathbf{X}_p'\mathbf{X}_p(\mathbf{X}_p'\mathbf{X}_p + \boldsymbol{\Gamma}_p \mathbf{K}_p \boldsymbol{\Gamma}_p')^{-1}\mathbf{S}_p^{-1}\boldsymbol{m}_p,$$
$$\mathrm{Cov}(\widehat{\boldsymbol{\theta}}_p(\mathbf{K}_p), \widehat{\theta}_0(\mathbf{K})) = \left\{\mathrm{Cov}(\widehat{\theta}_0(\mathbf{K}), \widehat{\boldsymbol{\theta}}_p(\mathbf{K}_p))\right\}'$$
$$= -\sigma^2 \mathbf{S}_p^{-1}(\mathbf{X}_p'\mathbf{X}_p + \boldsymbol{\Gamma}_p \mathbf{K}_p \boldsymbol{\Gamma}_p')^{-1}\mathbf{X}_p'\mathbf{X}_p(\mathbf{X}_p'\mathbf{X}_p + \boldsymbol{\Gamma}_p \mathbf{K}_p \boldsymbol{\Gamma}_p')^{-1}\mathbf{S}_p^{-1}\boldsymbol{m}_p,$$
$$\mathrm{V}(\widehat{\boldsymbol{\theta}}_p(\mathbf{K}_p))$$
$$= \sigma^2 \mathbf{S}_p^{-1}(\mathbf{X}_p'\mathbf{X}_p + \boldsymbol{\Gamma}_p \mathbf{K}_p \boldsymbol{\Gamma}_p')^{-1}\mathbf{X}_p'\mathbf{X}_p(\mathbf{X}_p'\mathbf{X}_p + \boldsymbol{\Gamma}_p \mathbf{K}_p \boldsymbol{\Gamma}_p')^{-1}\mathbf{S}_p^{-1}.$$

3.2. MEAN SQUARED ERROR CRITERIA OF REGRESSION ESTIMATORS 27

The MSE matrix is

$$\text{MSE}(\widehat{\boldsymbol{\theta}}(\mathbf{K})) = \text{V}(\widehat{\boldsymbol{\theta}}(\mathbf{K})) + \text{bias}(\widehat{\boldsymbol{\theta}}(\mathbf{K}))\text{bias}(\widehat{\boldsymbol{\theta}}(\mathbf{K}))'$$

$$= \begin{bmatrix} \text{V}(\widehat{\theta}_0(\mathbf{K})) & \text{Cov}(\widehat{\theta}_0(\mathbf{K}), \widehat{\boldsymbol{\theta}}_p(\mathbf{K}_p)) \\ \text{Cov}(\widehat{\boldsymbol{\theta}}_p(\mathbf{K}_p), \widehat{\theta}_0(\mathbf{K})) & \text{V}(\widehat{\boldsymbol{\theta}}_p(\mathbf{K}_p)) \end{bmatrix}$$

$$+ \begin{bmatrix} \text{bias}(\widehat{\theta}_0(\mathbf{K}))^2 & \text{bias}(\widehat{\theta}_0(\mathbf{K}))\text{bias}(\widehat{\boldsymbol{\theta}}_p(\mathbf{K}_p))' \\ \text{bias}(\widehat{\boldsymbol{\theta}}_p(\mathbf{K}_p))\text{bias}(\widehat{\theta}_0(\mathbf{K})) & \text{bias}(\widehat{\boldsymbol{\theta}}_p(\mathbf{K}_p))\text{bias}(\widehat{\boldsymbol{\theta}}_p(\mathbf{K}_p))' \end{bmatrix}$$

$$= \begin{bmatrix} \text{MSE}(\widehat{\theta}_0(\mathbf{K})) & \text{MCE}(\widehat{\theta}_0(\mathbf{K}), \widehat{\boldsymbol{\theta}}_p(\mathbf{K}_p)) \\ \text{MCE}(\widehat{\boldsymbol{\theta}}_p(\mathbf{K}_p), \widehat{\theta}_0(\mathbf{K})) & \text{MSE}(\widehat{\boldsymbol{\theta}}_p(\mathbf{K}_p)) \end{bmatrix}, \quad (3.24)$$

where

$$\text{MSE}(\widehat{\theta}_0(\mathbf{K})) = \text{V}(\widehat{\theta}_0(\mathbf{K})) + \text{bias}(\widehat{\theta}_0(\mathbf{K}))^2$$
$$= \sigma^2 \frac{n}{(n+k_0)^2} + \sigma^2 \boldsymbol{m}_p' \mathbf{S}_p^{-1} (\mathbf{X}_p'\mathbf{X}_p + \boldsymbol{\Gamma}_p \mathbf{K}_p \boldsymbol{\Gamma}_p')^{-1} \mathbf{X}_p' \mathbf{X}_p (\mathbf{X}_p'\mathbf{X}_p + \boldsymbol{\Gamma}_p \mathbf{K}_p \boldsymbol{\Gamma}_p')^{-1} \mathbf{S}_p^{-1} \boldsymbol{m}_p$$
$$+ \left(-\frac{k_0}{n+k_0} \beta_0 - \boldsymbol{m}_p' \mathbf{S}_p^{-1} \left\{ (\mathbf{X}_p'\mathbf{X}_p + \boldsymbol{\Gamma}_p \mathbf{K}_p \boldsymbol{\Gamma}_p')^{-1} \mathbf{X}_p' \mathbf{X}_p - \mathbf{I}_p \right\} \boldsymbol{\beta}_p \right)^2$$
$$= \sigma^2 \frac{n}{(n+k_0)^2} + \sigma^2 \boldsymbol{m}_p' \mathbf{S}_p^{-1} (\mathbf{X}_p'\mathbf{X}_p + \boldsymbol{\Gamma}_p \mathbf{K}_p \boldsymbol{\Gamma}_p')^{-1} \mathbf{X}_p' \mathbf{X}_p (\mathbf{X}_p'\mathbf{X}_p + \boldsymbol{\Gamma}_p \mathbf{K}_p \boldsymbol{\Gamma}_p')^{-1} \mathbf{S}_p^{-1} \boldsymbol{m}_p$$
$$+ \frac{k_0^2}{(n+k_0)^2} \beta_0^2 + 2 \frac{k_0}{n+k_0} \beta_0 \boldsymbol{m}_p' \mathbf{S}_p^{-1} \left\{ (\mathbf{X}_p'\mathbf{X}_p + \boldsymbol{\Gamma}_p \mathbf{K}_p \boldsymbol{\Gamma}_p')^{-1} \mathbf{X}_p' \mathbf{X}_p - \mathbf{I}_p \right\} \boldsymbol{\beta}_p$$
$$+ \left(\boldsymbol{m}_p' \mathbf{S}_p^{-1} \left\{ (\mathbf{X}_p'\mathbf{X}_p + \boldsymbol{\Gamma}_p \mathbf{K}_p \boldsymbol{\Gamma}_p')^{-1} \mathbf{X}_p' \mathbf{X}_p - \mathbf{I}_p \right\} \boldsymbol{\beta}_p \right)^2,$$

$$\text{MCE}(\widehat{\boldsymbol{\theta}}_p(\mathbf{K}_p), \widehat{\theta}_0(\mathbf{K})) = \text{Cov}(\widehat{\boldsymbol{\theta}}_p(\mathbf{K}_p), \widehat{\theta}_0(\mathbf{K})) + \text{bias}(\widehat{\boldsymbol{\theta}}_p(\mathbf{K}_p))\text{bias}(\widehat{\theta}_0(\mathbf{K}))$$
$$= -\sigma^2 \mathbf{S}_p^{-1} (\mathbf{X}_p'\mathbf{X}_p + \boldsymbol{\Gamma}_p \mathbf{K}_p \boldsymbol{\Gamma}_p')^{-1} \mathbf{X}_p' \mathbf{X}_p (\mathbf{X}_p'\mathbf{X}_p + \boldsymbol{\Gamma}_p \mathbf{K}_p \boldsymbol{\Gamma}_p')^{-1} \mathbf{S}_p^{-1} \boldsymbol{m}_p$$
$$+ \mathbf{S}_p^{-1} \left\{ (\mathbf{X}_p'\mathbf{X}_p + \boldsymbol{\Gamma}_p \mathbf{K}_p \boldsymbol{\Gamma}_p')^{-1} \mathbf{X}_p' \mathbf{X}_p - \mathbf{I}_p \right\} \boldsymbol{\beta}_p$$
$$\times \left\{ -\frac{k_0}{n+k_0} \beta_0 - \boldsymbol{m}_p' \mathbf{S}_p^{-1} \left\{ (\mathbf{X}_p'\mathbf{X}_p + \boldsymbol{\Gamma}_p \mathbf{K}_p \boldsymbol{\Gamma}_p')^{-1} \mathbf{X}_p' \mathbf{X}_p - \mathbf{I}_p \right\} \boldsymbol{\beta}_p \right\},$$

$$\text{MSE}(\widehat{\boldsymbol{\theta}}_p(\mathbf{K}_p)) = \text{V}(\widehat{\boldsymbol{\theta}}_p(\mathbf{K}_p)) + \text{bias}(\widehat{\boldsymbol{\theta}}_p(\mathbf{K}_p))\text{bias}(\widehat{\boldsymbol{\theta}}_p(\mathbf{K}_p))'$$
$$= \sigma^2 \mathbf{S}_p^{-1} (\mathbf{X}_p'\mathbf{X}_p + \boldsymbol{\Gamma}_p \mathbf{K}_p \boldsymbol{\Gamma}_p')^{-1} \mathbf{X}_p' \mathbf{X}_p (\mathbf{X}_p'\mathbf{X}_p + \boldsymbol{\Gamma}_p \mathbf{K}_p \boldsymbol{\Gamma}_p')^{-1} \mathbf{S}_p^{-1}$$
$$+ \mathbf{S}_p^{-1} \left\{ (\mathbf{X}_p'\mathbf{X}_p + \boldsymbol{\Gamma}_p \mathbf{K}_p \boldsymbol{\Gamma}_p')^{-1} \mathbf{X}_p' \mathbf{X}_p - \mathbf{I}_p \right\} \boldsymbol{\beta}_p$$
$$\times \boldsymbol{\beta}_p' \left\{ \mathbf{X}_p' \mathbf{X}_p (\mathbf{X}_p'\mathbf{X}_p + \boldsymbol{\Gamma}_p \mathbf{K}_p \boldsymbol{\Gamma}_p')^{-1} - \mathbf{I}_p \right\} \mathbf{S}_p^{-1}.$$

Consequently, the TMSE is

$$\text{TMSE}(\widehat{\boldsymbol{\theta}}(\mathbf{K})) = trace\ \text{MSE}(\widehat{\boldsymbol{\theta}}(\mathbf{K})) = \text{MSE}(\widehat{\theta}_0(\mathbf{K})) + trace\ \text{MSE}(\widehat{\boldsymbol{\theta}}_p(\mathbf{K}_p))$$

$$= \sigma^2 \frac{n}{(n+k_0)^2} + \frac{k_0^2}{(n+k_0)^2}\beta_0^2$$
$$+ 2\frac{k_0}{n+k_0}\beta_0 \mathbf{m}'_p \mathbf{S}_p^{-1} \left\{ (\mathbf{X}'_p\mathbf{X}_p + \boldsymbol{\Gamma}_p\mathbf{K}_p\boldsymbol{\Gamma}'_p)^{-1}\mathbf{X}'_p\mathbf{X}_p - \mathbf{I}_p \right\} \boldsymbol{\beta}_p$$
$$+ \sigma^2 trace\ (\mathbf{X}'_p\mathbf{X}_p + \boldsymbol{\Gamma}_p\mathbf{K}_p\boldsymbol{\Gamma}'_p)^{-1}\mathbf{X}'_p\mathbf{X}_p(\mathbf{X}'_p\mathbf{X}_p + \boldsymbol{\Gamma}_p\mathbf{K}_p\boldsymbol{\Gamma}'_p)^{-1}\left(\mathbf{S}_p^{-1}\mathbf{m}_p\mathbf{m}'_p\mathbf{S}_p^{-1} + \mathbf{S}_p^{-2}\right)$$
$$+ \boldsymbol{\beta}'_p \left\{ \mathbf{X}'_p\mathbf{X}_p(\mathbf{X}'_p\mathbf{X}_p + \boldsymbol{\Gamma}_p\mathbf{K}_p\boldsymbol{\Gamma}'_p)^{-1} - \mathbf{I}_p \right\}$$
$$\times \left(\mathbf{S}_p^{-1}\mathbf{m}_p\mathbf{m}'_p\mathbf{S}_p^{-1} + \mathbf{S}_p^{-2}\right) \left\{ (\mathbf{X}'_p\mathbf{X}_p + \boldsymbol{\Gamma}_p\mathbf{K}_p\boldsymbol{\Gamma}'_p)^{-1}\mathbf{X}'_p\mathbf{X}_p - \mathbf{I}_p \right\} \boldsymbol{\beta}_p$$
$$= \sigma^2 \frac{n}{(n+k_0)^2} + \frac{k_0^2}{(n+k_0)^2}\alpha_0^2 + 2\frac{k_0}{n+k_0}\alpha_0 \mathbf{m}'_p \mathbf{S}_p^{-1}\boldsymbol{\Gamma}_p \left\{ \boldsymbol{\Lambda}_p(\mathbf{K}_p)^{-1}\boldsymbol{\Lambda}_p - \mathbf{I}_p \right\} \boldsymbol{\alpha}_p$$
$$+ \sigma^2 trace\ \boldsymbol{\Lambda}_p(\mathbf{K}_p)^{-1}\boldsymbol{\Lambda}_p\boldsymbol{\Lambda}_p(\mathbf{K}_p)^{-1}\boldsymbol{\Gamma}'_p \left(\mathbf{S}_p^{-1}\mathbf{m}_p\mathbf{m}'_p\mathbf{S}_p^{-1} + \mathbf{S}_p^{-2}\right) \boldsymbol{\Gamma}_p$$
$$+ \boldsymbol{\alpha}'_p \left\{ \boldsymbol{\Lambda}_p\boldsymbol{\Lambda}_p(\mathbf{K}_p)^{-1} - \mathbf{I}_p \right\} \boldsymbol{\Gamma}'_p \left(\mathbf{S}_p^{-1}\mathbf{m}_p\mathbf{m}'_p\mathbf{S}_p^{-1} + \mathbf{S}_p^{-2}\right) \boldsymbol{\Gamma}_p \left\{ \boldsymbol{\Lambda}_p(\mathbf{K}_p)^{-1}\boldsymbol{\Lambda}_p - \mathbf{I}_p \right\} \boldsymbol{\alpha}_p.$$
(3.25)

3.2.4 PCR Estimator

Here, we discuss the MSE criteria of the PCR estimator. We first consider the following PCR estimator for $\boldsymbol{\beta}_p$:

$$\widehat{\boldsymbol{\beta}}_p(r,\cdot) = \boldsymbol{\Gamma}_r \boldsymbol{\Lambda}_r^{-1} \boldsymbol{\Gamma}'_r \mathbf{X}'_p \mathbf{Y}.$$

The following PCR estimator is biased.

$$\text{E}(\widehat{\boldsymbol{\beta}}_p(r,\cdot)) = \boldsymbol{\Gamma}_r \boldsymbol{\Lambda}_r^{-1} \boldsymbol{\Gamma}'_r \mathbf{X}'_p \text{E}(\mathbf{Y}) = \boldsymbol{\Gamma}_r \boldsymbol{\Lambda}_r^{-1} \boldsymbol{\Gamma}'_r \mathbf{X}'_p \mathbf{X}_p \boldsymbol{\beta}_p = \boldsymbol{\Gamma}_r \boldsymbol{\Gamma}'_r \boldsymbol{\beta}_p,$$

where we use the partitions

$$\boldsymbol{\Gamma}_p = [\boldsymbol{\Gamma}_r, \boldsymbol{\Gamma}_{p\backslash r}], \quad \boldsymbol{\Lambda}_p = \begin{bmatrix} \boldsymbol{\Lambda}_r & \mathbf{O} \\ \mathbf{O} & \boldsymbol{\Lambda}_{p\backslash r} \end{bmatrix},$$

in which $\mathbf{O}$ indicates the zero matrix, and we have used the equalities

$$\mathbf{X}'_p \mathbf{X}_p = \boldsymbol{\Gamma}_p \boldsymbol{\Lambda}_p \boldsymbol{\Gamma}'_p = [\boldsymbol{\Gamma}_r, \boldsymbol{\Gamma}_{p\backslash r}] \begin{bmatrix} \boldsymbol{\Lambda}_r & \mathbf{O} \\ \mathbf{O} & \boldsymbol{\Lambda}_{p\backslash r} \end{bmatrix} \begin{bmatrix} \boldsymbol{\Gamma}'_r \\ \boldsymbol{\Gamma}'_{p\backslash r} \end{bmatrix} = \boldsymbol{\Gamma}_r \boldsymbol{\Lambda}_r \boldsymbol{\Gamma}'_r + \boldsymbol{\Gamma}_{p\backslash r} \boldsymbol{\Lambda}_{p\backslash r} \boldsymbol{\Gamma}'_{p\backslash r},$$

$$\boldsymbol{\Gamma}'_r \boldsymbol{\Gamma}_r = \mathbf{I}_r, \quad \boldsymbol{\Gamma}'_r \boldsymbol{\Gamma}_{p\backslash r} = \mathbf{O}.$$

We see that the bias vector of $\widehat{\boldsymbol{\beta}}_p(r,\cdot)$ is not the zero vector as follows:

$$\text{bias}(\widehat{\boldsymbol{\beta}}_p(r,\cdot)) = \text{E}(\widehat{\boldsymbol{\beta}}_p(r,\cdot)) - \boldsymbol{\beta}_p = -(\mathbf{I}_p - \boldsymbol{\Gamma}_r\boldsymbol{\Gamma}'_r)\boldsymbol{\beta}_p = -\boldsymbol{\Gamma}_{p\backslash r}\boldsymbol{\Gamma}'_{p\backslash r}\boldsymbol{\beta}_p \neq \mathbf{0},$$

where we have used that

$$\mathbf{I}_p = \boldsymbol{\Gamma}_p \boldsymbol{\Gamma}'_p = [\boldsymbol{\Gamma}_r, \boldsymbol{\Gamma}_{p\backslash r}] \begin{bmatrix} \boldsymbol{\Gamma}'_r \\ \boldsymbol{\Gamma}'_{p\backslash r} \end{bmatrix} = \boldsymbol{\Gamma}_r \boldsymbol{\Gamma}'_r + \boldsymbol{\Gamma}_{p\backslash r} \boldsymbol{\Gamma}'_{p\backslash r}.$$

3.2. MEAN SQUARED ERROR CRITERIA OF REGRESSION ESTIMATORS

The variance-covariance matrix of the PCR estimator is

$$V(\widehat{\boldsymbol{\beta}}_p(r,\cdot)) = \boldsymbol{\Gamma}_r\boldsymbol{\Lambda}_r^{-1}\boldsymbol{\Gamma}_r'\mathbf{X}_p'E(\boldsymbol{Y})\mathbf{X}_p\boldsymbol{\Gamma}_r\boldsymbol{\Lambda}_r^{-1}\boldsymbol{\Gamma}_r'$$
$$= \sigma^2\boldsymbol{\Gamma}_r\boldsymbol{\Lambda}_r^{-1}\boldsymbol{\Gamma}_r'\mathbf{X}_p'\mathbf{X}_p\boldsymbol{\Gamma}_r\boldsymbol{\Lambda}_r^{-1}\boldsymbol{\Gamma}_r'$$
$$= \sigma^2\boldsymbol{\Gamma}_r\boldsymbol{\Lambda}_r^{-1}\boldsymbol{\Gamma}_r'.$$

Therefore, the MSE matrix is

$$\mathrm{MSE}(\widehat{\boldsymbol{\beta}}_p(r,\cdot)) = V(\widehat{\boldsymbol{\beta}}_p(r,\cdot)) + \mathrm{bias}(\widehat{\boldsymbol{\beta}}_p(r,\cdot))\mathrm{bias}(\widehat{\boldsymbol{\beta}}_p(r,\cdot))'$$
$$= \sigma^2\boldsymbol{\Gamma}_r\boldsymbol{\Lambda}_r^{-1}\boldsymbol{\Gamma}_r' + \boldsymbol{\Gamma}_{p\backslash r}\boldsymbol{\Gamma}_{p\backslash r}'\boldsymbol{\beta}_p\boldsymbol{\beta}_p'\boldsymbol{\Gamma}_{p\backslash r}\boldsymbol{\Gamma}_{p\backslash r}', \qquad (3.26)$$

and the TMSE is

$$\mathrm{TMSE}(\widehat{\boldsymbol{\beta}}_p(r,\cdot)) = trace\ \mathrm{MSE}(\widehat{\boldsymbol{\beta}}_p(r,\cdot))$$
$$= \sigma^2 trace\ \boldsymbol{\Gamma}_r\boldsymbol{\Lambda}_r^{-1}\boldsymbol{\Gamma}_r' + trace\ \boldsymbol{\Gamma}_{p\backslash r}\boldsymbol{\Gamma}_{p\backslash r}'\boldsymbol{\beta}_p\boldsymbol{\beta}_p'\boldsymbol{\Gamma}_{p\backslash r}\boldsymbol{\Gamma}_{p\backslash r}'$$
$$= \sigma^2 trace\ \boldsymbol{\Lambda}_r^{-1} + \boldsymbol{\alpha}_{p\backslash r}'\boldsymbol{\alpha}_{p\backslash r}$$
$$= \sigma^2 \sum_{i=1}^{r}\frac{1}{\lambda_i} + \sum_{i=r+1}^{p}\alpha_i^2, \qquad (3.27)$$

where

$$\boldsymbol{\Gamma}_p'\boldsymbol{\beta}_p = \begin{bmatrix} \boldsymbol{\Gamma}_r'\boldsymbol{\beta}_p \\ \boldsymbol{\Gamma}_{p\backslash r}'\boldsymbol{\beta}_p \end{bmatrix} = \begin{bmatrix} \boldsymbol{\alpha}_p \\ \boldsymbol{\alpha}_{p\backslash r} \end{bmatrix} = \boldsymbol{\alpha}_p.$$

Next, we consider the MSE criteria of the PCR estimator

$$\widehat{\boldsymbol{\beta}}(r,\cdot) = \begin{bmatrix} \overline{Y} \\ \boldsymbol{\Gamma}_r\boldsymbol{\Lambda}_r^{-1}\boldsymbol{\Gamma}_r'\mathbf{X}_p'Y \end{bmatrix} = \begin{bmatrix} \widehat{\beta}_0 \\ \widehat{\boldsymbol{\beta}}_p(r,\cdot) \end{bmatrix}.$$

Similar to the case $\widehat{\boldsymbol{\beta}}_p(r,\cdot)$,

$$\mathrm{bias}(\widehat{\boldsymbol{\beta}}(r,\cdot)) = \begin{bmatrix} 0 \\ -\boldsymbol{\Gamma}_{p\backslash r}\boldsymbol{\Gamma}_{p\backslash r}'\boldsymbol{\beta}_p \end{bmatrix} = \begin{bmatrix} \mathrm{bias}(\widehat{\beta}_0) \\ \mathrm{bias}(\widehat{\boldsymbol{\beta}}_p(r,\cdot)) \end{bmatrix},$$

$$V(\widehat{\boldsymbol{\beta}}(r,\cdot)) = \begin{bmatrix} \frac{\sigma^2}{n} & 0 \\ 0 & \sigma^2\boldsymbol{\Gamma}_r\boldsymbol{\Lambda}_r^{-1}\boldsymbol{\Gamma}_r' \end{bmatrix} = \begin{bmatrix} V(\widehat{\beta}_0) & \mathrm{Cov}(\widehat{\beta}_0,\widehat{\boldsymbol{\beta}}_p(r,\cdot)) \\ \mathrm{Cov}(\widehat{\boldsymbol{\beta}}_p(r,\cdot),\widehat{\beta}_0) & V(\widehat{\boldsymbol{\beta}}_p(r,\cdot)) \end{bmatrix}.$$

Therefore, the MSE matrix is

$$\mathrm{MSE}(\widehat{\boldsymbol{\beta}}(r,\cdot)) = V(\widehat{\boldsymbol{\beta}}(r,\cdot)) + \mathrm{bias}(\widehat{\boldsymbol{\beta}}(r,\cdot))\mathrm{bias}(\widehat{\boldsymbol{\beta}}(r,\cdot))'$$
$$= \begin{bmatrix} \frac{\sigma^2}{n} & 0 \\ 0 & \sigma^2\boldsymbol{\Gamma}_r\boldsymbol{\Lambda}_r^{-1}\boldsymbol{\Gamma}_r' \end{bmatrix} + \begin{bmatrix} 0 & 0 \\ 0 & \boldsymbol{\Gamma}_{p\backslash r}\boldsymbol{\Gamma}_{p\backslash r}'\boldsymbol{\beta}_p\boldsymbol{\beta}_p'\boldsymbol{\Gamma}_{p\backslash r}\boldsymbol{\Gamma}_{p\backslash r}' \end{bmatrix}$$
$$= \begin{bmatrix} \frac{\sigma^2}{n} & 0 \\ 0 & \sigma^2\boldsymbol{\Gamma}_r\boldsymbol{\Lambda}_r^{-1}\boldsymbol{\Gamma}_r' + \boldsymbol{\Gamma}_{p\backslash r}\boldsymbol{\Gamma}_{p\backslash r}'\boldsymbol{\beta}_p\boldsymbol{\beta}_p'\boldsymbol{\Gamma}_{p\backslash r}\boldsymbol{\Gamma}_{p\backslash r}' \end{bmatrix}$$
$$= \begin{bmatrix} \mathrm{MSE}(\widehat{\beta}_0) & 0 \\ 0 & \mathrm{MSE}(\widehat{\boldsymbol{\beta}}_p(r,\cdot)) \end{bmatrix}, \qquad (3.28)$$

and the TMSE is

$$\text{TMSE}(\widehat{\boldsymbol{\beta}}(r,\cdot)) = trace \begin{bmatrix} \text{MSE}(\widehat{\beta}_0) & 0 \\ 0 & \text{MSE}(\widehat{\boldsymbol{\beta}}_p(r,\cdot)) \end{bmatrix}$$

$$= \text{MSE}(\widehat{\beta}_0) + \text{TMSE}(\widehat{\boldsymbol{\beta}}_p(r,\cdot)) = \frac{\sigma^2}{n} + \left(\sigma^2 \sum_{i=1}^{r} \frac{1}{\lambda_i} + \sum_{i=r+1}^{p} \alpha_i^2\right). \quad (3.29)$$

Finally, we consider the MSE criteria of the PCR-type estimator

$$\widehat{\boldsymbol{\theta}}(r,\cdot) = \mathbf{T}^{-1}\widehat{\boldsymbol{\beta}}(r,\cdot) = \begin{bmatrix} 1 & -\boldsymbol{m}_p'\mathbf{S}_p^{-1} \\ 0 & \mathbf{S}_p^{-1} \end{bmatrix} \begin{bmatrix} \widehat{\beta}_0 \\ \widehat{\boldsymbol{\beta}}_p(r,\cdot) \end{bmatrix} = \begin{bmatrix} \widehat{\beta}_0 - \boldsymbol{m}_p'\mathbf{S}_p^{-1}\widehat{\boldsymbol{\beta}}_p(r,\cdot) \\ \mathbf{S}_p^{-1}\widehat{\boldsymbol{\beta}}_p(r,\cdot) \end{bmatrix}.$$

The bias vector and variance-covariance matrix are

$$\text{bias}(\widehat{\boldsymbol{\theta}}(r,\cdot)) = \text{bias}(\mathbf{T}^{-1}\widehat{\boldsymbol{\beta}}(r,\cdot)) = \mathbf{T}^{-1}\text{bias}(\widehat{\boldsymbol{\beta}}(r,\cdot))$$

$$= \begin{bmatrix} 1 & -\boldsymbol{m}_p'\mathbf{S}_p^{-1} \\ 0 & \mathbf{S}_p^{-1} \end{bmatrix} \begin{bmatrix} 0 \\ -\boldsymbol{\Gamma}_{p\backslash r}\boldsymbol{\Gamma}_{p\backslash r}'\boldsymbol{\beta}_p \end{bmatrix}$$

$$= \begin{bmatrix} \boldsymbol{m}_p'\mathbf{S}_p^{-1}\boldsymbol{\Gamma}_{p\backslash r}\boldsymbol{\Gamma}_{p\backslash r}'\boldsymbol{\beta}_p \\ -\mathbf{S}_p^{-1}\boldsymbol{\Gamma}_{p\backslash r}\boldsymbol{\Gamma}_{p\backslash r}'\boldsymbol{\beta}_p \end{bmatrix} = \begin{bmatrix} \text{bias}(\widehat{\theta}_0(r,\cdot)) \\ \text{bias}(\widehat{\boldsymbol{\theta}}_p(r,\cdot)) \end{bmatrix},$$

and

$$\text{V}(\widehat{\boldsymbol{\theta}}(r,\cdot)) = \text{V}(\mathbf{T}^{-1}\widehat{\boldsymbol{\beta}}(r,\cdot)) = \mathbf{T}^{-1}\text{V}(\widehat{\boldsymbol{\beta}}(r,\cdot))(\mathbf{T}^{-1})'$$

$$= \begin{bmatrix} 1 & -\boldsymbol{m}_p'\mathbf{S}_p^{-1} \\ 0 & \mathbf{S}_p^{-1} \end{bmatrix} \begin{bmatrix} \frac{\sigma^2}{n} & 0 \\ 0 & \sigma^2\boldsymbol{\Gamma}_r\boldsymbol{\Lambda}_r^{-1}\boldsymbol{\Gamma}_r' \end{bmatrix} \begin{bmatrix} 1 & 0 \\ -\mathbf{S}_p^{-1}\boldsymbol{m}_p & \mathbf{S}_p^{-1} \end{bmatrix}$$

$$= \begin{bmatrix} \frac{\sigma^2}{n} + \sigma^2\boldsymbol{m}_p'\mathbf{S}_p^{-1}\boldsymbol{\Gamma}_r\boldsymbol{\Lambda}_r^{-1}\boldsymbol{\Gamma}_r'\mathbf{S}_p^{-1}\boldsymbol{m}_p & -\sigma^2\boldsymbol{m}_p'\mathbf{S}_p^{-1}\boldsymbol{\Gamma}_r\boldsymbol{\Lambda}_r^{-1}\boldsymbol{\Gamma}_r' \\ -\sigma^2\mathbf{S}_p^{-1}\boldsymbol{\Gamma}_r\boldsymbol{\Lambda}_r^{-1}\boldsymbol{\Gamma}_r'\mathbf{S}_p^{-1}\boldsymbol{m}_p & \sigma^2\mathbf{S}_p^{-1}\boldsymbol{\Gamma}_r\boldsymbol{\Lambda}_r^{-1}\boldsymbol{\Gamma}_r'\mathbf{S}_p^{-1} \end{bmatrix}$$

$$= \begin{bmatrix} \text{V}(\widehat{\theta}_0(r,\cdot)) & \text{Cov}(\widehat{\theta}_0(r,\cdot),\widehat{\boldsymbol{\theta}}_p(r,\cdot)) \\ \text{Cov}(\widehat{\boldsymbol{\theta}}_p(r,\cdot),\widehat{\theta}_0(r,\cdot)) & \text{V}(\widehat{\boldsymbol{\theta}}_p(r,\cdot)) \end{bmatrix}.$$

Therefore, the MSE matrix is

$$\text{MSE}(\widehat{\boldsymbol{\theta}}(r,\cdot)) \quad (3.30)$$

$$= \text{V}(\widehat{\boldsymbol{\theta}}(r,\cdot)) + \text{bias}(\widehat{\boldsymbol{\theta}}(r,\cdot))\text{bias}(\widehat{\boldsymbol{\theta}}(r,\cdot))'$$

$$= \begin{bmatrix} \frac{\sigma^2}{n} + \sigma^2\boldsymbol{m}_p'\mathbf{S}_p^{-1}\boldsymbol{\Gamma}_r\boldsymbol{\Lambda}_r^{-1}\boldsymbol{\Gamma}_r'\mathbf{S}_p^{-1}\boldsymbol{m}_p & -\sigma^2\boldsymbol{m}_p'\mathbf{S}_p^{-1}\boldsymbol{\Gamma}_r\boldsymbol{\Lambda}_r^{-1}\boldsymbol{\Gamma}_r' \\ -\sigma^2\mathbf{S}_p^{-1}\boldsymbol{\Gamma}_r\boldsymbol{\Lambda}_r^{-1}\boldsymbol{\Gamma}_r'\mathbf{S}_p^{-1}\boldsymbol{m}_p & \sigma^2\mathbf{S}_p^{-1}\boldsymbol{\Gamma}_r\boldsymbol{\Lambda}_r^{-1}\boldsymbol{\Gamma}_r'\mathbf{S}_p^{-1} \end{bmatrix}$$

$$+ \begin{bmatrix} \boldsymbol{m}_p'\mathbf{S}_p^{-1}\boldsymbol{\Gamma}_{p\backslash r}\boldsymbol{\Gamma}_{p\backslash r}'\boldsymbol{\beta}_p\boldsymbol{\beta}_p'\boldsymbol{\Gamma}_{p\backslash r}\boldsymbol{\Gamma}_{p\backslash r}'\mathbf{S}_p^{-1}\boldsymbol{m}_p & -\boldsymbol{m}_p'\mathbf{S}_p^{-1}\boldsymbol{\Gamma}_{p\backslash r}\boldsymbol{\Gamma}_{p\backslash r}'\boldsymbol{\beta}_p\boldsymbol{\beta}_p'\mathbf{S}_p^{-1}\boldsymbol{\Gamma}_{p\backslash r}\boldsymbol{\Gamma}_{p\backslash r}'\boldsymbol{\beta}_p \\ -\mathbf{S}_p^{-1}\boldsymbol{\Gamma}_{p\backslash r}\boldsymbol{\Gamma}_{p\backslash r}'\boldsymbol{\beta}_p\boldsymbol{\beta}_p'\boldsymbol{\Gamma}_{p\backslash r}\boldsymbol{\Gamma}_{p\backslash r}'\mathbf{S}_p^{-1}\boldsymbol{m}_p & \mathbf{S}_p^{-1}\boldsymbol{\Gamma}_{p\backslash r}\boldsymbol{\Gamma}_{p\backslash r}'\boldsymbol{\beta}_p\boldsymbol{\beta}_p'\boldsymbol{\Gamma}_{p\backslash r}\boldsymbol{\Gamma}_{p\backslash r}'\mathbf{S}_p^{-1} \end{bmatrix}$$

$$= \begin{bmatrix} \text{MSE}(\widehat{\theta}_0(r,\cdot)) & \text{MCE}(\widehat{\theta}_0(r,\cdot),\widehat{\boldsymbol{\theta}}_p(r,\cdot)) \\ \text{MCE}(\widehat{\boldsymbol{\theta}}_p(r,\cdot),\widehat{\theta}_0(r,\cdot)) & \text{MSE}(\widehat{\boldsymbol{\theta}}_p(r,\cdot)) \end{bmatrix}, \quad (3.31)$$

3.2. MEAN SQUARED ERROR CRITERIA OF REGRESSION ESTIMATORS

and the TMSE is

$$\text{TMSE}(\widehat{\boldsymbol{\theta}}(r,\cdot)) = trace\ \text{MSE}(\widehat{\boldsymbol{\theta}}(r,\cdot)) = \text{MSE}(\widehat{\theta}_0(r,\cdot)) + trace\ \text{MSE}(\widehat{\boldsymbol{\theta}}_p(r,\cdot))$$
$$= \left(\frac{\sigma^2}{n} + \sigma^2 \boldsymbol{m}'_p \mathbf{S}_p^{-1}\boldsymbol{\Gamma}_r\boldsymbol{\Lambda}_r^{-1}\boldsymbol{\Gamma}'_r\mathbf{S}_p^{-1}\boldsymbol{m}_p + \boldsymbol{m}'_p\mathbf{S}_p^{-1}\boldsymbol{\Gamma}_{p\backslash r}\boldsymbol{\Gamma}'_{p\backslash r}\boldsymbol{\beta}_p\boldsymbol{\beta}'_p\boldsymbol{\Gamma}_{p\backslash r}\boldsymbol{\Gamma}'_{p\backslash r}\mathbf{S}_p^{-1}\boldsymbol{m}_p\right)$$
$$+ \left(\sigma^2 trace\ \mathbf{S}_p^{-1}\boldsymbol{\Gamma}_r\boldsymbol{\Lambda}_r^{-1}\boldsymbol{\Gamma}'_r\mathbf{S}_p^{-1} + trace\ \mathbf{S}_p^{-1}\boldsymbol{\Gamma}_{p\backslash r}\boldsymbol{\Gamma}'_{p\backslash r}\boldsymbol{\beta}_p\boldsymbol{\beta}'_p\boldsymbol{\Gamma}_{p\backslash r}\boldsymbol{\Gamma}'_{p\backslash r}\mathbf{S}_p^{-1}\right). \tag{3.32}$$

3.2.5 r-k Class Estimator

As a final case, we discuss the MSE criteria of the r-k class estimator. We start by considering the following r-k class estimator for $\boldsymbol{\beta}_p$:

$$\widehat{\boldsymbol{\beta}}_p(r,k) = \boldsymbol{\Gamma}_r\boldsymbol{\Lambda}_r(k)^{-1}\boldsymbol{\Gamma}'_r\mathbf{X}'_p\boldsymbol{Y}.$$

The following r-k estimator is biased:

$$\text{E}(\widehat{\boldsymbol{\beta}}_p(r,\cdot)) = \boldsymbol{\Gamma}_r\boldsymbol{\Lambda}_r(k)^{-1}\boldsymbol{\Gamma}'_r\mathbf{X}'_p\text{E}(\boldsymbol{Y}) = \boldsymbol{\Gamma}_r\boldsymbol{\Lambda}_r(k)^{-1}\boldsymbol{\Gamma}'_r\mathbf{X}'_p\mathbf{X}_p\boldsymbol{\beta}_p$$
$$= \boldsymbol{\Gamma}_r\boldsymbol{\Lambda}_r(k)^{-1}\boldsymbol{\Lambda}_r\boldsymbol{\Gamma}'_r\boldsymbol{\beta}_p.$$

Thus, the bias vector of $\widehat{\boldsymbol{\beta}}_p$ is not the zero vector:

$$\text{bias}(\widehat{\boldsymbol{\beta}}_p(r,k)) = \text{E}(\widehat{\boldsymbol{\beta}}_p) - \boldsymbol{\beta}_p = -(\mathbf{I}_p - \boldsymbol{\Gamma}_r\boldsymbol{\Lambda}_r(k)^{-1}\boldsymbol{\Lambda}_r\boldsymbol{\Gamma}'_r)\boldsymbol{\beta}_p \neq \mathbf{0}.$$

The variance-covariance matrix of the r-k class estimator is

$$\text{V}(\widehat{\boldsymbol{\beta}}_p(r,k)) = \boldsymbol{\Gamma}_r\boldsymbol{\Lambda}_r(k)^{-1}\boldsymbol{\Gamma}'_r\mathbf{X}'_p\text{E}(\boldsymbol{Y})\mathbf{X}_p\boldsymbol{\Gamma}_r\boldsymbol{\Lambda}_r(k)^{-1}\boldsymbol{\Gamma}'_r$$
$$= \sigma^2\boldsymbol{\Gamma}_r\boldsymbol{\Lambda}_r(k)^{-1}\boldsymbol{\Gamma}'_r\mathbf{X}'_p\mathbf{X}_p\boldsymbol{\Gamma}_r\boldsymbol{\Lambda}_r(k)^{-1}\boldsymbol{\Gamma}'_r$$
$$= \sigma^2\boldsymbol{\Gamma}_r\boldsymbol{\Lambda}_r(k)^{-1}\boldsymbol{\Lambda}_r\boldsymbol{\Lambda}_r(k)^{-1}\boldsymbol{\Gamma}'_r.$$

Therefore, the MSE matrix is

$$\text{MSE}(\widehat{\boldsymbol{\beta}}_p(r,k)) = \text{V}(\widehat{\boldsymbol{\beta}}_p(r,k)) + \text{bias}(\widehat{\boldsymbol{\beta}}_p(r,k))\text{bias}(\widehat{\boldsymbol{\beta}}_p(r,k))'$$
$$= \sigma^2\boldsymbol{\Gamma}_r\boldsymbol{\Lambda}_r(k)^{-1}\boldsymbol{\Lambda}_r\boldsymbol{\Lambda}_r(k)^{-1}\boldsymbol{\Gamma}'_r$$
$$+ (\mathbf{I}_p - \boldsymbol{\Gamma}_r\boldsymbol{\Lambda}_r(k)^{-1}\boldsymbol{\Lambda}_r\boldsymbol{\Gamma}'_r)\boldsymbol{\beta}_p\boldsymbol{\beta}'_p(\mathbf{I}_p - \boldsymbol{\Gamma}_r\boldsymbol{\Lambda}_r(k)^{-1}\boldsymbol{\Lambda}_r\boldsymbol{\Gamma}'_r), \tag{3.33}$$

and the TMSE is

$$\text{TMSE}(\widehat{\boldsymbol{\beta}}_p(r,k)) = trace\ \text{MSE}(\widehat{\boldsymbol{\beta}}_p(r,k))$$
$$= \sigma^2 trace\ \boldsymbol{\Gamma}_r\boldsymbol{\Lambda}_r(k)^{-1}\boldsymbol{\Lambda}_r\boldsymbol{\Lambda}_r(k)^{-1}\boldsymbol{\Gamma}'_r$$
$$+ trace\ (\mathbf{I}_p - \boldsymbol{\Gamma}_r\boldsymbol{\Lambda}_r(k)^{-1}\boldsymbol{\Lambda}_r\boldsymbol{\Gamma}'_r)\boldsymbol{\beta}_p\boldsymbol{\beta}'_p(\mathbf{I}_p - \boldsymbol{\Gamma}_r\boldsymbol{\Lambda}_r(k)^{-1}\boldsymbol{\Lambda}_r\boldsymbol{\Gamma}'_r)$$
$$= \sigma^2 trace\ \boldsymbol{\Lambda}_r(k)^{-1}\boldsymbol{\Lambda}_r\boldsymbol{\Lambda}_r(k)^{-1}$$
$$+ trace\ \boldsymbol{\beta}'_p(\mathbf{I}_p - \boldsymbol{\Gamma}_r\boldsymbol{\Lambda}_r(k)^{-1}\boldsymbol{\Lambda}_r\boldsymbol{\Gamma}'_r)(\mathbf{I}_p - \boldsymbol{\Gamma}_r\boldsymbol{\Lambda}_r(k)^{-1}\boldsymbol{\Lambda}_r\boldsymbol{\Gamma}'_r)\boldsymbol{\beta}_p$$
$$= \sigma^2\sum_{i=1}^{r}\frac{\lambda_i}{(\lambda_i+k)^2} + \left(\sum_{i=1}^{r}\frac{k^2\alpha_i^2}{(\lambda_i+k)^2} + \sum_{i=r+1}^{p}\alpha_i^2\right). \tag{3.34}$$

Next, we consider the MSE criteria of the r-k estimator

$$\widehat{\boldsymbol{\beta}}(r,k_0) = \begin{bmatrix} \frac{n}{n+k_0}\overline{Y} \\ \boldsymbol{\Gamma}_r\boldsymbol{\Lambda}_r(k)^{-1}\boldsymbol{\Gamma}_r'\mathbf{X}_p'\mathbf{Y} \end{bmatrix} = \begin{bmatrix} \widehat{\beta}_0(k_0) \\ \widehat{\boldsymbol{\beta}}_p(r,k) \end{bmatrix}.$$

Similar to the case $\widehat{\boldsymbol{\beta}}_p(r,k)$,

$$\text{bias}(\widehat{\boldsymbol{\beta}}(r,k)) = \begin{bmatrix} -\frac{k_0}{n+k_0}\beta_0 \\ -(\mathbf{I}_p - \boldsymbol{\Gamma}_r\boldsymbol{\Lambda}_r(k)^{-1}\boldsymbol{\Lambda}_r\boldsymbol{\Gamma}_r')\boldsymbol{\beta}_p \end{bmatrix} = \begin{bmatrix} \text{bias}(\widehat{\beta}_0(k_0)) \\ \text{bias}(\widehat{\boldsymbol{\beta}}_p(r,k)) \end{bmatrix},$$

$$V(\widehat{\boldsymbol{\beta}}(r,k_0)) = \begin{bmatrix} \sigma^2\frac{n}{(n+k_0)^2} & 0 \\ 0 & \sigma^2\boldsymbol{\Gamma}_r\boldsymbol{\Lambda}_r(k)^{-1}\boldsymbol{\Lambda}_r\boldsymbol{\Lambda}_r(k)^{-1}\boldsymbol{\Gamma}_r' \end{bmatrix}$$
$$= \begin{bmatrix} V(\widehat{\beta}_0(k_0)) & \text{Cov}(\widehat{\beta}_0(k_0),\widehat{\boldsymbol{\beta}}_p(r,k)) \\ \text{Cov}(\widehat{\boldsymbol{\beta}}_p(r,k),\widehat{\beta}_0(k_0)) & V(\widehat{\boldsymbol{\beta}}_p(r,k)) \end{bmatrix}.$$

Therefore, the MSE matrix is

$$\text{MSE}(\widehat{\boldsymbol{\beta}}(r,k_0)) = V(\widehat{\boldsymbol{\beta}}(r,k_0)) + \text{bias}(\widehat{\boldsymbol{\beta}}(r,k_0))\text{bias}(\widehat{\boldsymbol{\beta}}(r,k_0))'$$
$$= \begin{bmatrix} V(\widehat{\beta}_0(k_0)) & \text{Cov}(\widehat{\beta}_0(k_0),\widehat{\boldsymbol{\beta}}_p(r,k)) \\ \text{Cov}(\widehat{\boldsymbol{\beta}}_p(r,k),\widehat{\beta}_0(k_0)) & V(\widehat{\boldsymbol{\beta}}_p(r,k)) \end{bmatrix}$$
$$+ \begin{bmatrix} \text{bias}(\widehat{\beta}_0(k_0))^2 & \text{bias}(\widehat{\beta}_0(k_0))\text{bias}(\widehat{\boldsymbol{\beta}}_p(r,k))' \\ \text{bias}(\widehat{\boldsymbol{\beta}}_p(r,k))\text{bias}(\widehat{\beta}_0(k_0)) & \text{bias}(\widehat{\boldsymbol{\beta}}_p(r,k))\text{bias}(\widehat{\boldsymbol{\beta}}_p(r,k))' \end{bmatrix}$$
$$= \begin{bmatrix} \text{MSE}(\widehat{\beta}_0(k_0)) & \text{MCE}(\widehat{\beta}_0(k_0),\widehat{\boldsymbol{\beta}}_p(r,k)) \\ \text{MCE}(\widehat{\boldsymbol{\beta}}_p(r,k),\widehat{\beta}_0(k_0)) & \text{MSE}(\widehat{\boldsymbol{\beta}}_p(r,k)) \end{bmatrix}, \quad (3.35)$$

where

$$\text{MSE}(\widehat{\beta}_0(k_0)) = V(\widehat{\beta}_0(k_0)) + \text{bias}(\widehat{\beta}_0(k_0))^2$$
$$= \sigma^2\frac{n}{(n+k_0)^2} + \frac{k_0^2}{(n+k_0)^2}\beta_0^2,$$
$$\text{MCE}(\widehat{\boldsymbol{\beta}}_p(r,k),\widehat{\beta}_0(k_0)) = \text{Cov}(\widehat{\boldsymbol{\beta}}_p(r,k),\widehat{\beta}_0(k_0)) + \text{bias}(\widehat{\boldsymbol{\beta}}_p(r,k))\text{bias}(\widehat{\beta}_0(k_0))$$
$$= \frac{k_0}{(n+k_0)^2}\beta_0(\mathbf{I}_p - \boldsymbol{\Gamma}_r\boldsymbol{\Lambda}_r(k)^{-1}\boldsymbol{\Lambda}_r\boldsymbol{\Gamma}_r')\boldsymbol{\beta}_p$$
$$= \text{MCE}(\widehat{\beta}_0(k_0),\widehat{\boldsymbol{\beta}}_p(r,k))',$$

and the TMSE is

$$\text{TMSE}(\widehat{\boldsymbol{\beta}}(r,k_0)) = \text{MSE}(\widehat{\beta}_0(k_0)) + \text{TMSE}(\widehat{\boldsymbol{\beta}}_p(r,k))$$
$$= \left(\sigma^2\frac{n}{(n+k_0)^2} + \frac{k_0^2}{(n+k_0)^2}\beta_0^2\right) + \left\{\sigma^2\sum_{i=1}^{r}\frac{\lambda_i}{(\lambda_i+k)^2} + \left(\sum_{i=1}^{r}\frac{k^2\alpha_i^2}{(\lambda_i+k)^2} + \sum_{i=r+1}^{p}\alpha_i^2\right)\right\}$$
(3.36)

3.2. MEAN SQUARED ERROR CRITERIA OF REGRESSION ESTIMATORS

Finally, we consider the MSE criteria of the r-k-class-type estimator

$$\widehat{\boldsymbol{\theta}}(r,k_0) = \mathbf{T}^{-1}\widehat{\boldsymbol{\beta}}(r,k_0) = \begin{bmatrix} 1 & -\boldsymbol{m}'_p\mathbf{S}_p^{-1} \\ 0 & \mathbf{S}_p^{-1} \end{bmatrix}\begin{bmatrix} \widehat{\beta}_0(k_0) \\ \widehat{\boldsymbol{\beta}}_p(r,k) \end{bmatrix}$$

$$= \begin{bmatrix} \widehat{\beta}_0(k_0) - \boldsymbol{m}'_p\mathbf{S}_p^{-1}\widehat{\boldsymbol{\beta}}_p(r,k) \\ \mathbf{S}_p^{-1}\widehat{\boldsymbol{\beta}}_p(r,k) \end{bmatrix}.$$

The bias vector and variance-covariance matrix are

$$\text{bias}(\widehat{\boldsymbol{\theta}}(r,k_0)) = \mathbf{T}^{-1}\text{bias}(\widehat{\boldsymbol{\beta}}(r,k_0))$$

$$= \begin{bmatrix} 1 & -\boldsymbol{m}'_p\mathbf{S}_p^{-1} \\ 0 & \mathbf{S}_p^{-1} \end{bmatrix}\begin{bmatrix} -\frac{k_0}{n+k_0}\beta_0 \\ -(\mathbf{I}_p - \boldsymbol{\Gamma}_r\boldsymbol{\Lambda}_r(k)^{-1}\boldsymbol{\Lambda}_r\boldsymbol{\Gamma}'_r)\boldsymbol{\beta}_p \end{bmatrix}$$

$$= \begin{bmatrix} -\frac{k_0}{n+k_0}\beta_0 + \boldsymbol{m}'_p\mathbf{S}_p^{-1}(\mathbf{I}_p - \boldsymbol{\Gamma}_r\boldsymbol{\Lambda}_r(k)^{-1}\boldsymbol{\Lambda}_r\boldsymbol{\Gamma}'_r)\boldsymbol{\beta}_p \\ -\mathbf{S}_p^{-1}(\mathbf{I}_p - \boldsymbol{\Gamma}_r\boldsymbol{\Lambda}_r(k)^{-1}\boldsymbol{\Lambda}_r\boldsymbol{\Gamma}'_r)\boldsymbol{\beta}_p \end{bmatrix}$$

$$= \begin{bmatrix} \text{bias}(\widehat{\theta}_0(r,k_0)) \\ \text{bias}(\widehat{\boldsymbol{\theta}}_p(r,k)) \end{bmatrix}.$$

$$V(\widehat{\boldsymbol{\theta}}(r,k_0)) = \mathbf{T}^{-1}V(\widehat{\boldsymbol{\beta}}(r,k_0))(\mathbf{T}^{-1})'$$

$$= \begin{bmatrix} 1 & -\boldsymbol{m}'_p\mathbf{S}_p^{-1} \\ 0 & \mathbf{S}_p^{-1} \end{bmatrix}\begin{bmatrix} \sigma^2\frac{n}{(n+k_0)^2} & 0 \\ 0 & \sigma^2\boldsymbol{\Gamma}_r\boldsymbol{\Lambda}_r(k)^{-1}\boldsymbol{\Lambda}_r\boldsymbol{\Lambda}_r(k)^{-1}\boldsymbol{\Gamma}'_r \end{bmatrix}\begin{bmatrix} 1 & 0 \\ -\mathbf{S}_p^{-1}\boldsymbol{m}_p & \mathbf{S}_p^{-1} \end{bmatrix}$$

$$= \begin{bmatrix} V(\widehat{\theta}_0(r,k_0)) & \text{Cov}(\widehat{\theta}_0(r,k_0),\widehat{\boldsymbol{\theta}}_p(r,k)) \\ \text{Cov}(\widehat{\boldsymbol{\theta}}_p(r,k),\widehat{\theta}_0(r,k_0)) & V(\widehat{\boldsymbol{\theta}}_p(r,k)) \end{bmatrix},$$

where

$$V(\widehat{\theta}_0(r,k_0)) = \sigma^2\frac{n}{(n+k_0)^2} + \sigma^2\boldsymbol{m}'_p\mathbf{S}_p^{-1}\boldsymbol{\Gamma}_r\boldsymbol{\Lambda}_r(k)^{-1}\boldsymbol{\Lambda}_r\boldsymbol{\Lambda}_r(k)^{-1}\boldsymbol{\Gamma}'_r\mathbf{S}_p^{-1}\boldsymbol{m}_p,$$

$$\text{Cov}(\widehat{\boldsymbol{\theta}}_p(r,k),\widehat{\theta}_0(r,k_0)) = -\sigma^2\mathbf{S}_p^{-1}\boldsymbol{\Gamma}_r\boldsymbol{\Lambda}_r(k)^{-1}\boldsymbol{\Lambda}_r\boldsymbol{\Lambda}_r(k)^{-1}\boldsymbol{\Gamma}'_r\mathbf{S}_p^{-1}\boldsymbol{m}_p$$

$$= \text{Cov}(\widehat{\theta}_0(r,k_0),\widehat{\boldsymbol{\theta}}_p(r,k))',$$

$$V(\widehat{\boldsymbol{\theta}}_p(r,k)) = \sigma^2\mathbf{S}_p^{-1}\boldsymbol{\Gamma}_r\boldsymbol{\Lambda}_r(k)^{-1}\boldsymbol{\Lambda}_r\boldsymbol{\Lambda}_r(k)^{-1}\boldsymbol{\Gamma}'_r\mathbf{S}_p^{-1}.$$

Therefore, the MSE matrix is

$$\text{MSE}(\widehat{\boldsymbol{\theta}}(r,k_0)) = V(\widehat{\boldsymbol{\theta}}(r,k_0)) + \text{bias}(\widehat{\boldsymbol{\theta}}(r,k_0))\text{bias}(\widehat{\boldsymbol{\theta}}(r,k_0))'$$

$$= \begin{bmatrix} V(\widehat{\theta}_0(r,k_0)) & \text{Cov}(\widehat{\theta}_0(r,k_0),\widehat{\boldsymbol{\theta}}_p(r,k)) \\ \text{Cov}(\widehat{\boldsymbol{\theta}}_p(r,k),\widehat{\theta}_0(r,k_0)) & V(\widehat{\boldsymbol{\theta}}_p(r,k)) \end{bmatrix}$$

$$+ \begin{bmatrix} \text{bias}(\widehat{\theta}_0(r,k_0))^2 & \text{bias}(\widehat{\theta}_0(r,k_0))\text{bias}(\widehat{\boldsymbol{\theta}}_p(r,k))' \\ \text{bias}(\widehat{\boldsymbol{\theta}}_p(r,k))\text{bias}(\widehat{\theta}_0(r,k_0)) & \text{bias}(\widehat{\boldsymbol{\theta}}_p(r,k))\text{bias}(\widehat{\boldsymbol{\theta}}_p(r,k))' \end{bmatrix}$$

$$= \begin{bmatrix} \text{MSE}(\widehat{\theta}_0(r,k_0)) & \text{MCE}(\widehat{\theta}_0(r,k_0),\widehat{\boldsymbol{\theta}}_p(r,k)) \\ \text{MCE}(\widehat{\boldsymbol{\theta}}_p(r,k),\widehat{\theta}_0(r,k_0)) & \text{MSE}(\widehat{\boldsymbol{\theta}}_p(r,k)) \end{bmatrix},$$

(3.37)

where

$$\begin{aligned}
\text{MSE}(\widehat{\theta}_0(r,k_0)) &= \text{V}(\widehat{\theta}_0(r,k_0)) + \text{bias}(\widehat{\theta}_0(r,k_0))^2 \\
&= \sigma^2\frac{n}{(n+k_0)^2} + \sigma^2 m'_p \mathbf{S}_p^{-1}\boldsymbol{\Gamma}_r\boldsymbol{\Lambda}_r(k)^{-1}\boldsymbol{\Lambda}_r\boldsymbol{\Lambda}_r(k)^{-1}\boldsymbol{\Gamma}'_r\mathbf{S}_p^{-1}m_p \\
&\quad + \left(-\frac{k_0}{n+k_0}\beta_0 + m'_p\mathbf{S}_p^{-1}(\mathbf{I}_p - \boldsymbol{\Gamma}_r\boldsymbol{\Lambda}_r(k)^{-1}\boldsymbol{\Lambda}_r\boldsymbol{\Gamma}'_r)\boldsymbol{\beta}_p\right)^2 \\
&= \sigma^2\frac{n}{(n+k_0)^2} + \sigma^2 m'_p \mathbf{S}_p^{-1}\boldsymbol{\Gamma}_r\boldsymbol{\Lambda}_r(k)^{-1}\boldsymbol{\Lambda}_r\boldsymbol{\Lambda}_r(k)^{-1}\boldsymbol{\Gamma}'_r\mathbf{S}_p^{-1}m_p \\
&\quad + \frac{k_0^2}{(n+k_0)^2}\beta_0^2 - 2\frac{k_0}{n+k_0}\beta_0 m'_p\mathbf{S}_p^{-1}(\mathbf{I}_p - \boldsymbol{\Gamma}_r\boldsymbol{\Lambda}_r(k)^{-1}\boldsymbol{\Lambda}_r\boldsymbol{\Gamma}'_r)\boldsymbol{\beta}_p \\
&\quad + \left(m'_p\mathbf{S}_p^{-1}(\mathbf{I}_p - \boldsymbol{\Gamma}_r\boldsymbol{\Lambda}_r(k)^{-1}\boldsymbol{\Lambda}_r\boldsymbol{\Gamma}'_r)\boldsymbol{\beta}_p\right)^2,
\end{aligned}$$

$$\begin{aligned}
\text{MCE}(\widehat{\boldsymbol{\theta}}_p(r,k),\widehat{\theta}_0(r,k_0)) &= \text{Cov}(\widehat{\boldsymbol{\theta}}_p(r,k),\widehat{\theta}_0(r,k_0)) + \text{bias}(\widehat{\boldsymbol{\theta}}_p(r,k))\text{bias}(\widehat{\theta}_0(r,k_0)) \\
&= -\sigma^2\mathbf{S}_p^{-1}\boldsymbol{\Gamma}_r\boldsymbol{\Lambda}_r(k)^{-1}\boldsymbol{\Lambda}_r\boldsymbol{\Lambda}_r(k)^{-1}\boldsymbol{\Gamma}'_r\mathbf{S}_p^{-1}m_p \\
&\quad - \mathbf{S}_p^{-1}(\mathbf{I}_p - \boldsymbol{\Gamma}_r\boldsymbol{\Lambda}_r(k)^{-1}\boldsymbol{\Lambda}_r\boldsymbol{\Gamma}'_r)\boldsymbol{\beta}_p \\
&\quad \times \left(-\frac{k_0}{n+k_0}\beta_0 + m'_p\mathbf{S}_p^{-1}(\mathbf{I}_p - \boldsymbol{\Gamma}_r\boldsymbol{\Lambda}_r(k)^{-1}\boldsymbol{\Lambda}_r\boldsymbol{\Gamma}'_r)\boldsymbol{\beta}_p\right),
\end{aligned}$$

$$\begin{aligned}
\text{MSE}(\widehat{\boldsymbol{\theta}}_p(r,k)) &= \text{V}(\widehat{\boldsymbol{\theta}}_p(r,k)) + \text{bias}(\widehat{\boldsymbol{\theta}}_p(r,k))\text{bias}(\widehat{\boldsymbol{\theta}}_p(r,k))' \\
&= \sigma^2\mathbf{S}_p^{-1}\boldsymbol{\Gamma}_r\boldsymbol{\Lambda}_r(k)^{-1}\boldsymbol{\Lambda}_r\boldsymbol{\Lambda}_r(k)^{-1}\boldsymbol{\Gamma}'_r\mathbf{S}_p^{-1} \\
&\quad + \mathbf{S}_p^{-1}(\mathbf{I}_p - \boldsymbol{\Gamma}_r\boldsymbol{\Lambda}_r(k)^{-1}\boldsymbol{\Lambda}_r\boldsymbol{\Gamma}'_r)\boldsymbol{\beta}_p\boldsymbol{\beta}'_p(\mathbf{I}_p - \boldsymbol{\Gamma}_r\boldsymbol{\Lambda}_r(k)^{-1}\boldsymbol{\Lambda}_r\boldsymbol{\Gamma}'_r)\mathbf{S}_p^{-1}.
\end{aligned}$$

Consequently, the TMSE is

$$\begin{aligned}
&\text{TMSE}(\widehat{\boldsymbol{\theta}}(r,k_0)) \\
&= \text{trace } \text{MSE}(\widehat{\boldsymbol{\theta}}(r,k_0)) = \text{MSE}(\widehat{\theta}_0(r,k_0)) + \text{trace } \text{MSE}(\widehat{\boldsymbol{\theta}}_p(r,k)) \\
&= \sigma^2\frac{n}{(n+k_0)^2} + \frac{k_0^2}{(n+k_0)^2}\beta_0^2 - 2\frac{k_0}{n+k_0}\beta_0 m'_p\mathbf{S}_p^{-1}(\mathbf{I}_p - \boldsymbol{\Gamma}_r\boldsymbol{\Lambda}_r(k)^{-1}\boldsymbol{\Lambda}_r\boldsymbol{\Gamma}'_r)\boldsymbol{\beta}_p \\
&\quad + \sigma^2\text{trace } \boldsymbol{\Lambda}_r(k)^{-1}\boldsymbol{\Lambda}_r\boldsymbol{\Lambda}_r(k)^{-1}\boldsymbol{\Gamma}'_r\left(\mathbf{S}_p^{-1}m_p m'_p\mathbf{S}_p^{-1} + \mathbf{S}_p^{-2}\right)\boldsymbol{\Gamma}_r \\
&\quad + \boldsymbol{\beta}'_p(\mathbf{I}_p - \boldsymbol{\Gamma}_r\boldsymbol{\Lambda}_r(k)^{-1}\boldsymbol{\Lambda}_r\boldsymbol{\Gamma}'_r)\left(\mathbf{S}_p^{-1}m_p m'_p\mathbf{S}_p^{-1} + \mathbf{S}_p^{-2}\right)(\mathbf{I}_p - \boldsymbol{\Gamma}_r\boldsymbol{\Lambda}_r(k)^{-1}\boldsymbol{\Lambda}_r\boldsymbol{\Gamma}'_r)\boldsymbol{\beta}_p \\
&= \sigma^2\frac{n}{(n+k_0)^2} + \frac{k_0^2}{(n+k_0)^2}\beta_0^2 + 2\frac{k_0}{n+k_0}\beta_0 m'_p\mathbf{S}_p^{-1}(\boldsymbol{\Gamma}_r\boldsymbol{\Lambda}_r(k)^{-1}\boldsymbol{\Lambda}_r\boldsymbol{\Gamma}'_r - \mathbf{I}_p)\boldsymbol{\beta}_p \\
&\quad + \sigma^2\text{trace } \boldsymbol{\Lambda}_r(k)^{-1}\boldsymbol{\Lambda}_r\boldsymbol{\Lambda}_r(k)^{-1}\boldsymbol{\Gamma}'_r\left(\mathbf{S}_p^{-1}m_p m'_p\mathbf{S}_p^{-1} + \mathbf{S}_p^{-2}\right)\boldsymbol{\Gamma}_r \\
&\quad + \boldsymbol{\beta}'_p(\mathbf{I}_p - \boldsymbol{\Gamma}_r\boldsymbol{\Lambda}_r(k)^{-1}\boldsymbol{\Lambda}_r\boldsymbol{\Gamma}'_r)\left(\mathbf{S}_p^{-1}m_p m'_p\mathbf{S}_p^{-1} + \mathbf{S}_p^{-2}\right)(\mathbf{I}_p - \boldsymbol{\Gamma}_r\boldsymbol{\Lambda}_r(k)^{-1}\boldsymbol{\Lambda}_r\boldsymbol{\Gamma}'_r)\boldsymbol{\beta}_p.
\end{aligned}$$
(3.38)

3.3 TMSE Comparisons among Regression Estimators

In this section, we consider several comparisons of the TMSE among the regression estimators given in Section 3.2. Let us first summarize the relevant important results of several papers. First of all, Hoerl and Kennard (1970a) reported that there always exists a positive value of k such that the TMSE of the ORR estimator is always less than that of the OLS estimator as follows:

$$\text{TMSE}(\widehat{\boldsymbol{\beta}}_p(\cdot, k)) < \text{TMSE}(\widehat{\boldsymbol{\beta}}_p).$$

(See Theorem 4.3, which is called the *existence theorem* in Hoerl and Kennard (1970a).) Next, Marquart (1970) gave a sufficient condition for the TMSE of the PCR estimator to be less than that of the OLS estimator. (See Theorem 15 in Marquart (1970).) Baye and Parker (1984) showed that the TMSE of the r-k class estimator is less than that of the PCR estimator under the same r but different (optimal) k value. Nomura and Ohkubo (1985) showed the superiority of the r-k class estimators with adaptive r, k over the OLS and ORR estimators. Jimichi and Inagaki (1996) rearranged the conditions treated in the above works and compared the TMSE of the estimators.

The conditions are denoted (C1) to (C4) as follows:

$$(\text{C1}) : 0 < k \leq \frac{2\sigma^2}{\|\boldsymbol{\beta}_p\|^2}$$

$$(\text{C2}) : \forall i \in N_{p \backslash r}, \quad \frac{\sigma^2}{\lambda_i} - \alpha_i^2 > 0$$

$$(\text{C3}) : 0 < k \leq \frac{2\sigma^2}{\|\boldsymbol{\alpha}_r\|^2}$$

$$(\text{C4}) : 0 \leq k < \min_{i \in N_{p \backslash r}} \left\{ \frac{\lambda_i}{2\alpha_i^2} \left(\frac{\sigma^2}{\lambda_i} - \alpha_i^2 \right) \right\}$$

where r is fixed in $\{1, \ldots, p-1\}$. The TMSE relationships presented in Jimichi (2005) are summarized in Figure 3.1. (See Figure 4.1 in Jimichi (2005).)

Remark 3.2. *Note that conditions (C1) to (C4) are only sufficient conditions for the relationships shown in Figure* 3.1 *holding. The conditions with respect to k are especially conservative.*

Remark 3.3. *Jimichi and Inagaki (1996) also treated the following model with the matrix of concomitant variable* $\mathbf{C}$:

$$\begin{aligned} \boldsymbol{Y}^* &= \mathbf{X}\boldsymbol{\beta} + \mathbf{C}\boldsymbol{\delta} + \boldsymbol{\varepsilon} \\ &= \beta_0 \mathbf{1} + \mathbf{X}_p \boldsymbol{\beta}_p + \mathbf{C}\boldsymbol{\delta} + \boldsymbol{\varepsilon}. \end{aligned}$$

$$\begin{array}{ccc}
& (\text{C1}) & \\
\text{TMSE}\left(\widehat{\boldsymbol{\beta}}_p(\cdot,k)\right) & < & \text{TMSE}\left(\widehat{\boldsymbol{\beta}}_p\right) \\
(\text{C4}) \quad \vee & & \vee \quad (\text{C2}) \\
\text{TMSE}\left(\widehat{\boldsymbol{\beta}}_p(r,k)\right) & < & \text{TMSE}\left(\widehat{\boldsymbol{\beta}}_p(r,\cdot)\right) \\
& (\text{C3}) &
\end{array}$$

Figure 3.1: TMSE relationships for $\widehat{\boldsymbol{\beta}}_p(r,k), \widehat{\boldsymbol{\beta}}_p(r,\cdot), \widehat{\boldsymbol{\beta}}_p(\cdot,k), \widehat{\boldsymbol{\beta}}_p$

They considered the following estimators for $\boldsymbol{\beta}_p$ in this model:

$$\widehat{\boldsymbol{\beta}}_p^* := (\mathbf{X}_p'\mathbf{X}_p)^{-1}\mathbf{X}_p'\boldsymbol{Y}^*,$$
$$\widehat{\boldsymbol{\beta}}_p^*(\cdot,k) := (\mathbf{X}_p'\mathbf{X}_p + k\mathbf{I}_p)^{-1}\mathbf{X}_p'\boldsymbol{Y}^*,$$
$$\widehat{\boldsymbol{\beta}}_p^*(r,\cdot) := \boldsymbol{\Gamma}_r\boldsymbol{\Lambda}_r^{-1}\boldsymbol{\Gamma}_r'\mathbf{X}_p'\boldsymbol{Y}^*,$$
$$\widehat{\boldsymbol{\beta}}_p^*(r,k) := \boldsymbol{\Gamma}_r(\boldsymbol{\Lambda}_r + k\mathbf{I}_r)^{-1}\boldsymbol{\Gamma}_r'\mathbf{X}_p'\boldsymbol{Y}^*.$$

They found the relationships given in Figure 3.2 for the TMSEs of these estimators under the following conditions $(C1)^*$ to $(C4)^*$:

$(C1^*) : (C1)$ and $\alpha_i \eta_i \geq 0$, for any $i \in N_p$
$(C2^*) \equiv (C2)$
$(C3^*) : (C3)$ and $\alpha_i \eta_i \geq 0$, for any $i \in N_r$
$(C4^*) : 0 \leq k < \min\limits_{i \in N_{p\backslash r}} \left\{ \dfrac{\lambda_i}{|\alpha_i|}\left(\sqrt{\dfrac{\sigma^2}{\lambda_i}} - |\alpha_i|\right)\right\}$

where

$$\boldsymbol{\eta}_p = [\eta_1, \cdots, \eta_r, \eta_{r+1}, \cdots, \eta_p]' = \begin{bmatrix} \boldsymbol{\eta}_r \\ \boldsymbol{\eta}_{p\backslash r} \end{bmatrix} := \begin{bmatrix} \boldsymbol{\Gamma}_r'\mathbf{X}_p'\mathbf{C}\boldsymbol{\delta} \\ \boldsymbol{\Gamma}_{p\backslash r}'\mathbf{X}_p'\mathbf{C}\boldsymbol{\delta} \end{bmatrix} = \boldsymbol{\Gamma}_p'\mathbf{X}_p'\mathbf{C}\boldsymbol{\delta}.$$

See Theorems 3.1, 3.2 and Corollaries 3.1, 3.2 in Jimichi and Inagaki (1996).

3.3. TMSE COMPARISONS AMONG REGRESSION ESTIMATORS 37

$$\begin{array}{ccc}
& (\text{C1}^*) & \\
\text{TMSE}\left(\widehat{\boldsymbol{\beta}}_p^*(\cdot,k)\right) & < & \text{TMSE}\left(\widehat{\boldsymbol{\beta}}_p^*\right) \\
(\text{C4}^*) \quad \vee & & \vee \quad (\text{C2}^*) \\
\text{TMSE}\left(\widehat{\boldsymbol{\beta}}_p^*(r,k)\right) & < & \text{TMSE}\left(\widehat{\boldsymbol{\beta}}_p^*(r,\cdot)\right) \\
& (\text{C3}^*) &
\end{array}$$

Figure 3.2: TMSE relationships for $\widehat{\boldsymbol{\beta}}_p^*(r,k)$, $\widehat{\boldsymbol{\beta}}_p^*(r,\cdot)$, $\widehat{\boldsymbol{\beta}}_p^*(\cdot,k)$, $\widehat{\boldsymbol{\beta}}_p^*$

Although the above results relate to the estimators for $\boldsymbol{\beta}_p$, estimation of $\boldsymbol{\theta}$ by using the shrinkage regression-type estimators is important in applications. Jimichi and Inagaki (1993) gave the following theorem with respect to the ORR-type estimators (2.13), (2.14), (2.15) in Section 2.5:

Theorem 3.1 (Jimichi and Inagaki (1993)). *For a fixed value of $k \geq 0$,*

$(i) \quad g(k) \leq -\alpha_0^2 \quad \Rightarrow \quad TMSE(\widehat{\boldsymbol{\theta}}(\cdot,k_0)) < TMSE(\widehat{\boldsymbol{\theta}}(\cdot,k)) < TMSE(\widehat{\boldsymbol{\theta}}(\cdot,0))$,
$\qquad\qquad\qquad\qquad\qquad$ *for $k_0 > k$*

$(ii) \quad -\alpha_0^2 < g(k) < \dfrac{\sigma^2}{n} \quad \Rightarrow \quad \begin{cases} TMSE(\widehat{\boldsymbol{\theta}}(\cdot,k_0(k))) < TMSE(\widehat{\boldsymbol{\theta}}(\cdot,0)) \\ TMSE(\widehat{\boldsymbol{\theta}}(\cdot,k_0(k))) \leq TMSE(\widehat{\boldsymbol{\theta}}(\cdot,k)) \end{cases}$

$(iii) \quad g(k) \geq \dfrac{\sigma^2}{n} \quad \Rightarrow \quad \begin{cases} TMSE(\widehat{\boldsymbol{\theta}}(\cdot,0)) < TMSE(\widehat{\boldsymbol{\theta}}(\cdot,k_0)), \text{ for } k_0 > 0 \\ TMSE(\widehat{\boldsymbol{\theta}}(\cdot,0)) < TMSE(\widehat{\boldsymbol{\theta}}(\cdot,k)) \end{cases}$

where

$$g(k) := \alpha_0 \boldsymbol{m}_p' \mathbf{S}_p^{-1} \boldsymbol{\Gamma}_p (\boldsymbol{\Lambda}_p(k)^{-1}\boldsymbol{\Lambda}_p - \mathbf{I}_p)\boldsymbol{\alpha}_p, \quad k_0(k) := \frac{n}{\alpha_0^2 + g(k)}\left(\frac{\sigma^2}{n} - g(k)\right).$$

Proof. We regard the TMSE of $\widehat{\boldsymbol{\theta}}(\cdot,k_0)$ given as (3.18) as a function of k_0 for a fixed k and investigate its variation by differentiating it. We set

$\rho(k_0) := \text{TMSE}(\widehat{\boldsymbol{\theta}}(\cdot,k_0))$

$= \sigma^2 \dfrac{n}{(n+k_0)^2} + \dfrac{k_0^2}{(n+k_0)^2}\alpha_0^2 + 2\dfrac{k_0}{n+k_0}\alpha_0 \boldsymbol{m}_p' \mathbf{S}_p^{-1} \boldsymbol{\Gamma}_p \left\{\boldsymbol{\Lambda}_p(k)^{-1}\boldsymbol{\Lambda}_p - \mathbf{I}_p\right\}\boldsymbol{\alpha}_p$

$\quad + \sigma^2 \text{trace } \boldsymbol{\Lambda}_p(k)^{-1}\boldsymbol{\Lambda}_p\boldsymbol{\Lambda}_p(k)^{-1}\boldsymbol{\Gamma}_p' \left(\mathbf{S}_p^{-1}\boldsymbol{m}_p\boldsymbol{m}_p'\mathbf{S}_p^{-1} + \mathbf{S}_p^{-2}\right)\boldsymbol{\Gamma}_p$

$\quad + \boldsymbol{\alpha}_p' \left\{\boldsymbol{\Lambda}_p\boldsymbol{\Lambda}_p(k)^{-1} - \mathbf{I}_p\right\}\boldsymbol{\Gamma}_p' \left(\mathbf{S}_p^{-1}\boldsymbol{m}_p\boldsymbol{m}_p'\mathbf{S}_p^{-1} + \mathbf{S}_p^{-2}\right)\boldsymbol{\Gamma}_p \left\{\boldsymbol{\Lambda}_p(k)^{-1}\boldsymbol{\Lambda}_p - \mathbf{I}_p\right\}\boldsymbol{\alpha}_p,$

and the derivative of $\rho(k_0)$ with respect to k_0 is

$$\begin{aligned}
\rho'(k_0) &= \frac{d}{dk_0}\rho(k_0) \\
&= -\frac{2n\sigma^2}{(n+k_0)^3} + \frac{2n\alpha_0^2 k_0}{(n+k_0)^3} + \frac{2ng(k)}{(n+k_0)^2} \\
&= \frac{2n}{(n+k_0)^3}(\alpha_0^2 + g(k))\left\{k_0 - \frac{n}{\alpha_0^2 + g(k)}\left(\frac{\sigma^2}{n} - g(k)\right)\right\} \\
&= \frac{2n}{(n+k_0)^3}(\alpha_0^2 + g(k))(k_0 - k_0(k)).
\end{aligned}$$

If $g(k) \leq -\alpha_0^2$, then $\rho'(k_0) < 0$ for $k_0 > 0$, and thus, $\rho(k_0)$ is monotonically decreasing. If $-\alpha_0^2 < g(k) < \frac{\sigma^2}{n}$, then

$$\rho'(k_0) \begin{cases} < 0, & \text{if } 0 \leq k_0 < k_0(k), \\ = 0, & \text{if } k_0 = k_0(k), \\ > 0, & \text{if } k_0(k) < k_0, \end{cases}$$

and thus $\rho(k_0)$ is minimized at $k_0(k)$. If $g(k) \geq \frac{\sigma^2}{n}$, then $\rho'(k_0) > 0$ for $k_0 > 0$, and thus $\rho(k_0)$ is monotonically increasing. The proof is complete. □

Theorem 3.1 can be extended to the cases of the GRR-type estimator $\widehat{\boldsymbol{\theta}}(k_0, \mathbf{K}_p) := \widehat{\boldsymbol{\theta}}(\mathbf{K})$ and the r-k-class-type estimator $\widehat{\boldsymbol{\theta}}(r, k)$.

Theorem 3.2. *For fixed values of* $k_1, \cdots, k_p \geq 0$,

$$\begin{aligned}
(i) &\quad g(\mathbf{K}_p) \leq -\alpha_0^2 &&\Rightarrow TMSE(\widehat{\boldsymbol{\theta}}(k_0, \mathbf{K}_p)) < TMSE(\widehat{\boldsymbol{\theta}}(0, \mathbf{K}_p)), \text{ for } k_0 > 0 \\
(ii) &\quad -\alpha_0^2 < g(\mathbf{K}_p) < \frac{\sigma^2}{n} &&\Rightarrow TMSE(\widehat{\boldsymbol{\theta}}(k_0(\mathbf{K}_p), \mathbf{K}_p)) < TMSE(\widehat{\boldsymbol{\theta}}(0, \mathbf{K}_p)) \\
(iii) &\quad g(\mathbf{K}_p) \geq \frac{\sigma^2}{n} &&\Rightarrow TMSE(\widehat{\boldsymbol{\theta}}(0, \mathbf{K}_p)) < TMSE(\widehat{\boldsymbol{\theta}}(k_0, \mathbf{K}_p)), \text{ for } k_0 > 0
\end{aligned}$$

where

$$g(\mathbf{K}_p) := \alpha_0 \mathbf{m}_p' \mathbf{S}_p^{-1} \boldsymbol{\Gamma}_p (\boldsymbol{\Lambda}_p(\mathbf{K}_p))^{-1} \boldsymbol{\Lambda}_p - \mathbf{I}_p) \boldsymbol{\alpha}_p, \quad k_0(\mathbf{K}_p) := \frac{n}{\alpha_0^2 + g(\mathbf{K}_p)}\left(\frac{\sigma^2}{n} - g(\mathbf{K}_p)\right)$$

Theorem 3.3. *For fixed values of* $r \in N_p$ *and* $k \geq 0$,

$$\begin{aligned}
(i) &\quad g(r,k) \leq -\alpha_0^2 &&\Rightarrow TMSE(\widehat{\boldsymbol{\theta}}(r,k_0)) < TMSE(\widehat{\boldsymbol{\theta}}(r,k)) < TMSE(\widehat{\boldsymbol{\theta}}(r,0)), \\
& && \quad \text{for } k_0 > k \\
(ii) &\quad -\alpha_0^2 < g(r,k) < \frac{\sigma^2}{n} &&\Rightarrow \begin{cases} TMSE(\widehat{\boldsymbol{\theta}}(r,k_0(r,k))) < TMSE(\widehat{\boldsymbol{\theta}}(r,0)) \\ TMSE(\widehat{\boldsymbol{\theta}}(r,k_0(r,k))) \leq TMSE(\widehat{\boldsymbol{\theta}}(r,k)) \end{cases} \\
(iii) &\quad g(r,k) \geq \frac{\sigma^2}{n} &&\Rightarrow \begin{cases} TMSE(\widehat{\boldsymbol{\theta}}(r,0)) < TMSE(\widehat{\boldsymbol{\theta}}(r,k_0), \text{ for } k_0 > 0 \\ TMSE(\widehat{\boldsymbol{\theta}}(r,0)) < TMSE(\widehat{\boldsymbol{\theta}}(r,k)) \end{cases}
\end{aligned}$$

where

$$g(r,k) := \beta_0 m_p' S_p^{-1}(\Gamma_r \Lambda_r(k)^{-1} \Lambda_r \Gamma_r' - I_p)\beta_p$$
$$= \alpha_0 m_p' S_p^{-1}(\Gamma_r(\Lambda_r(k)^{-1}\Lambda_r - I_r)\alpha_r - \Gamma_{p\setminus r}\alpha_{p\setminus r}),$$
$$k_0(r,k) := \frac{n}{\alpha_0^2 + g(r,k)}\left(\frac{\sigma^2}{n} - g(r,k)\right).$$

The proofs of Theorems 3.2, 3.3 are similar to that of Theorem 3.1.

3.4 Basic Methods for Choosing Number of Principal Components and Ridge Coefficients

When the shrinkage regression estimators are used in practice, the number of principal component r or the ridge coefficients $k, k_0, k_1, \ldots, k_p$ must be chosen. In this section, we consider several basic methods for choosing them.

3.4.1 Number of Principal Component

The basic method for choosing the number of principal components of the PCR estimator $\widehat{\beta}_p(r, \cdot)$ and the r-k class estimator $\widehat{\beta}_p(r, k)$ is derived from condition (C2) in Section 3.3:

$$r^* := \min\left\{r \mid \tau_i^2 < 1, \forall i \in N_{p\setminus r}\right\} = \min\left\{r \mid \tau_{r+1}^2 < 1, \cdots, \tau_p^2 < 1\right\}, \quad (3.39)$$

where

$$\tau_i := \frac{\alpha_i}{\sigma/\sqrt{\lambda_i}}. \quad (3.40)$$

Note that

$$\frac{\sigma^2}{\lambda_i} - \alpha_i^2 > 0 \iff \tau_i^2 < 1.$$

3.4.2 Ridge Coefficients

There has been various work related to choosing the ridge coefficient k of the ORR estimator $\widehat{\beta}_p(\cdot, k)$ (e.g. Horel *et al.* (1975), Lawless and Wang (1976), Dempster *et al.* (1977), Gibbons (1981), and Jimichi and Inagaki (1993)). The basic method [1] used in all of them is based on minimizing the

---

[1] There are many ideas about how to choose the ridge coefficient k. Most of these are based on (1) *ridge trace* by Hoerl and Kennard (1970b), (2) *total mean squared error of prediction* (TMSEP) or *prediction sum of squares* (PRESS) by Allen (1972, 1974) or C_L *statistics* by Mallows (1973), or (3) *Bayesian methods* by Lindley and Smith (1971) and Rolph (1976).

TMSE:

$$\rho(k) := \text{TMSE}(\widehat{\boldsymbol{\beta}}_p(\cdot, k))$$
$$= \sigma^2 \sum_{i=1}^{p} \frac{\lambda_i}{(\lambda_i + k)^2} + \sum_{i=1}^{p} \frac{k^2 \alpha_i^2}{(\lambda_i + k)^2} \longrightarrow \min_{k>0}. \quad (3.41)$$

In general, a value of k (say k^*) which minimizes $\rho(k)$ is a solution to the equation

$$\frac{d\rho(k)}{dk} = 2 \sum_{i=1}^{p} \frac{\lambda_i(\alpha_i^2 k - \sigma^2)}{(\lambda_i + k)^3} = 0. \quad (3.42)$$

Note that no explicit solution to (3.42) exists except in some special cases. (See Baldwin and Hoerl (1978).) Thus, we need to numerically find a solution.

Similarly, the method for choosing the k_i of the GRR estimator $\widehat{\boldsymbol{\beta}}_p(\mathbf{K}_p)$ is minimizing

$$\rho(k_1, \ldots, k_p) := \text{TMSE}(\widehat{\boldsymbol{\beta}}_p(\mathbf{K}_p))$$
$$= \sigma^2 \sum_{i=1}^{p} \frac{\lambda_i}{(\lambda_i + k_i)^2} + \sum_{i=1}^{p} \frac{k_i^2 \alpha_i^2}{(\lambda_i + k_i)^2}$$
$$= \sum_{i=1}^{p} \rho_i(k_i) \longrightarrow \min_{\substack{k_i > 0, \\ i=1,\ldots,p}}, \quad (3.43)$$

where

$$\rho_i(k_i) := \frac{\sigma^2 \lambda_i}{(\lambda_i + k_i)^2} + \frac{k_i^2 \alpha_i^2}{(\lambda_i + k_i)^2}$$

is the MSE of the GRR estimator $\widehat{\alpha}_i(k_i)$ for α_i. (See (2.4) and also Figure 3.3.) The values of k_i ($i = 1, \ldots, p$) which minimize $\rho(k_1, \ldots, k_p)$ are the solutions to the equations

$$\frac{\partial \rho(k_1, \ldots, k_p)}{\partial k_i} = \frac{\partial \rho_i(k_i)}{\partial k_i} = 2 \frac{\lambda_i(\alpha_i^2 k_i - \sigma^2)}{(\lambda_i + k_i)^3} = 0, \quad i = 1, \ldots, p, \quad (3.44)$$

which are explicitly

$$k_i = \frac{\sigma^2}{\alpha_i^2} =: k_i^*, \quad i = 1, \ldots, p, \quad (3.45)$$

under the condition $\alpha_i \neq 0$. That is, $\widehat{\boldsymbol{\beta}}_p(\mathbf{K}_p^*)$ is optimum:

$$\text{TMSE}(\widehat{\boldsymbol{\beta}}_p(\mathbf{K}_p^*)) \leq \text{TMSE}(\widehat{\boldsymbol{\beta}}_p(\mathbf{K}_p)), \quad \forall k_i > 0, i = 1, \ldots, p, \quad (3.46)$$

where $\mathbf{K}_p^* := \text{diag}(k_1^*, \ldots, k_p^*)$. (See Hoerl and Kennard (1970a) and Hoerl et al. (1975).)

3.4. BASIC METHODS FOR CHOOSING NUMBER OF PC AND RIDGE COEF. 41

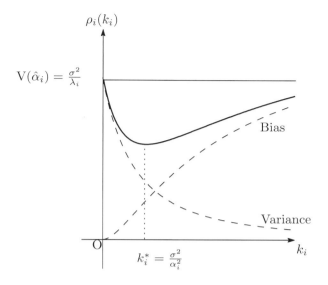

Figure 3.3: MSE of the GRR estimator $\widehat{\alpha}_i(k_i)$ for α_i.

Note that if $k_1 = \cdots = k_p = k$, then

$$\text{TMSE}(\widehat{\boldsymbol{\beta}}_p(\mathbf{K}_p)) = \rho(k_1, \ldots, k_p) = \rho(k, \cdots, k) = \rho(k) = \text{TMSE}(\widehat{\boldsymbol{\beta}}_p(k)),$$

and

$$\rho(k_1^*, \ldots, k_p^*) = \min_{\substack{k_i > 0, \\ i=1,\ldots,p}} \rho(k_1, \ldots, k_p)$$

$$\leq \min_{\substack{k_i > 0, \\ i=1,\ldots,p, \\ k_1 = \cdots = k_p = k}} \rho(k_1, \ldots, k_p) = \min_{k>0} \rho(k).$$

That is, for all $k > 0$,

$$\text{TMSE}(\widehat{\boldsymbol{\beta}}_p(\mathbf{K}_p^*)) \leq \text{TMSE}(\widehat{\boldsymbol{\beta}}_p(k)). \tag{3.47}$$

Remark 3.4. *Hoerl et al. (1975) proposed the following choice of the ridge coefficient k of the ORR estimator $\widehat{\boldsymbol{\beta}}_p(\cdot, k)$:*

$$k_h := \frac{p\sigma^2}{\|\boldsymbol{\beta}_p\|^2}. \tag{3.48}$$

This choice is related to the upper bounds in conditions (C1) and (C3). Note that if $\alpha_1^2 = \cdots = \alpha_p^2$, then the TMSE $\rho(k)$ of the ORR estimator $\widehat{\boldsymbol{\beta}}_p(\cdot, k)$ is minimized at $k = k_h$, which is the harmonic mean of k_i^*, $i = 1, \ldots, p$:

$$k_h = \left(\frac{1}{p} \sum_{i=1}^p \frac{1}{k_i^*}\right)^{-1}.$$

The method for choosing k of the r-k class estimator $\widehat{\boldsymbol{\beta}}_p(r,k)$ is as follows:

$$\rho(r,k) := \text{TMSE}(\widehat{\boldsymbol{\beta}}_p(r,k))$$
$$= \sigma^2 \sum_{i=1}^{r} \frac{\lambda_i}{(\lambda_i+k)^2} + \sum_{i=1}^{r} \frac{k^2 \alpha_i^2}{(\lambda_i+k)^2} + \sum_{i=r+1}^{p} \alpha_i^2 \longrightarrow \min_{k>0}. \quad (3.49)$$

A value of k (say k^*) which minimizes $\rho(r,k)$ is a solution to the equation

$$\frac{d\rho(r,k)}{dk} = 2 \sum_{i=1}^{r} \frac{\lambda_i(\alpha_i^2 k - \sigma^2)}{(\lambda_i+k)^3} = 0. \quad (3.50)$$

Note that no explicit solution to (3.50) exists.

The basic method for choosing the ridge coefficient k_0 of the ORR-type estimator $\widehat{\boldsymbol{\theta}}(\cdot, k_0)$ is derived from Theorem 3.1 as follows:

$$k_0^* = \begin{cases} \text{any value greater than } k, & \text{if } g(k) \leq -\alpha_0^2, \\ k_0(k), & \text{if } -\alpha_0^2 < g(k) < \dfrac{\sigma^2}{n}, \\ 0, & \text{if } g(k) \geq \dfrac{\sigma^2}{n}. \end{cases} \quad (3.51)$$

See also Jimichi and Inagaki (1993).

Similarly, the method for choosing the ridge coefficient k_0 of the GRR-type estimator $\widehat{\boldsymbol{\theta}}(k_0, \mathbf{K}_p)$ is derived from Theorem 3.2 as follows:

$$k_0^* = \begin{cases} \text{any value greater than } 0, & \text{if } g(\mathbf{K}_p) \leq -\alpha_0^2, \\ k_0(\mathbf{K}_p), & \text{if } -\alpha_0^2 < g(\mathbf{K}_p) < \dfrac{\sigma^2}{n}, \\ 0, & \text{if } g(\mathbf{K}_p) \geq \dfrac{\sigma^2}{n}. \end{cases} \quad (3.52)$$

Finally, the method for choosing the ridge coefficient k_0 of the r-k-class-type estimator $\widehat{\boldsymbol{\theta}}(r, k_0)$ is derived from Theorem 3.3 as follows:

$$k_0^* = \begin{cases} \text{any value greater than } k, & \text{if } g(r,k) \leq -\alpha_0^2, \\ k_0(r,k), & \text{if } -\alpha_0^2 < g(r,k) < \dfrac{\sigma^2}{n}, \\ 0, & \text{if } g(r,k) \geq \dfrac{\sigma^2}{n}. \end{cases} \quad (3.53)$$

Remark 3.5. *The methods for choosing the number of principal components r and the ridge coefficients k, $k_0, k_1, \ldots, k_p$ in Section 3.4 depend on unknown parameters. Thus, they cannot be directly used. For example, if we take $k_i = k_i^* (= \sigma^2/\alpha_i^2)$, then the GRR estimator $\widehat{\boldsymbol{\beta}}_p(\mathbf{K}_p^*)$ is optimum. (See (3.46).) However, the estimator is not feasible because k_i^* depends on unknown parameters σ^2, α_i^2. In the next chapter, we will discuss several feasible shrinkage regression estimators.*

Chapter 4
Feasible Shrinkage Regression Estimators

In this chapter, we consider several concrete algorithms for choosing the number of principal components r and the ridge coefficients $k, k_0, k_1, \ldots, k_p$ based on the basic methods in Section 3.4. If r or $k, k_0, k_1, \ldots, k_p$ chosen by the following algorithms are plugged into the appropriate parts of shrinkage regression estimators, then we will call the resulting estimators *feasible* shrinkage regression estimators.

4.1 Feasible ORR Estimator

Dempster *et al.* (1977) gave an algorithm for choosing the k of the ORR estimator $\widehat{\boldsymbol{\beta}}_p(\cdot, k)$ which has been much appreciated [1]. (See Gibbons (1981).) Their algorithm is based on the method of minimizing the TMSE $\rho(k)$ (see (3.42)) and has been implemented by evaluating

$$\left|\frac{\widehat{\rho}(k)'}{2}\right| = \left|\sum_{i=1}^{p} \frac{\lambda_i(\widehat{\alpha}_i(k)^2 k - \widehat{\sigma}^2)}{(\lambda_i + k)^3}\right| \tag{4.1}$$

for a mesh of k values and selecting that value of k (say $\widehat{k}^*$) associated with the observed minimum. $\widehat{\sigma}^2$ is the estimate for σ^2 defined in (1.14), and $\widehat{\alpha}_i(k)$ is the ith element of $\widehat{\boldsymbol{\alpha}}_p(\cdot, k) := \boldsymbol{\Gamma}'_p \widehat{\boldsymbol{\beta}}_p(\cdot, k)$. By plugging $\widehat{k}^*$ into the ORR estimator $\widehat{\boldsymbol{\beta}}_p(\cdot, k)$, we have the following feasible ORR estimator:

$$\widehat{\boldsymbol{\beta}}_p(\cdot, \widehat{k}^*) := (\mathbf{X}'_p \mathbf{X}_p + \widehat{k}^* \mathbf{I}_p)^{-1} \mathbf{X}'_p \mathbf{Y}. \tag{4.2}$$

An algorithm for the ridge coefficient k_0 of the ridge-type estimator $\widehat{\boldsymbol{\theta}}(\cdot, k_0)$ is given by (3.51) as follows:

$$\widehat{k}_0^* = \begin{cases} \text{any value greater than } \widehat{k}, & \text{if } \widehat{g}(\widehat{k}^*) \leq -\widehat{\alpha}_0^2, \\ \widehat{k}_0^*(\widehat{k}^*), & \text{if } -\widehat{\alpha}_0^2 < \widehat{g}(\widehat{k}^*) < \dfrac{\widehat{\sigma}^2}{n}, \\ 0, & \text{if } \widehat{g}(\widehat{k}^*) \geq \dfrac{\widehat{\sigma}^2}{n}. \end{cases} \tag{4.3}$$

---

[1] Note that many algorithms based on the TMSE have been proposed for choosing k (e.g. Hoerl *et al.* (1975), Lee (1987), Lee and Campbell (1985), and Jimichi and Inagaki (1993)).

where $\widehat{\boldsymbol{\alpha}} = [\widehat{\alpha}_0, \widehat{\boldsymbol{\alpha}}_p']'$ is the OLS estimator for $\boldsymbol{\alpha}$ in the canonical form (M_A) defined in (1.13) and

$$\widehat{g}(\widehat{k}^*) := \widehat{\alpha}_0 \boldsymbol{m}_p' \mathbf{S}_p^{-1} \boldsymbol{\Gamma}_p (\boldsymbol{\Lambda}_p(\widehat{k}^*)^{-1} \boldsymbol{\Lambda}_p - \mathbf{I}_p) \widehat{\boldsymbol{\alpha}}_p, \quad \widehat{k}_0^*(\widehat{k}^*) := \frac{n}{\widehat{\alpha}_0^2 + \widehat{g}(\widehat{k}^*)} \left(\frac{\widehat{\sigma}^2}{n} - \widehat{g}(\widehat{k}^*) \right).$$

We have the following feasible ORR-type estimator by plugging $\widehat{k}_0^*$ into the ORR-type estimator $\widehat{\boldsymbol{\theta}}_p(\cdot, k_0)$:

$$\widehat{\boldsymbol{\theta}}(\cdot, \widehat{k}_0^*) = \begin{cases} \widehat{\boldsymbol{\theta}}(\cdot, \widehat{k}_0^*), (\widehat{k}_0 > \widehat{k}^*) & \text{if } \widehat{g}(\widehat{k}^*) \leq -\widehat{\alpha}_0^2, \\ \widehat{\boldsymbol{\theta}}(\cdot, \widehat{k}_0^*(\widehat{k}^*)), & \text{if } -\widehat{\alpha}_0^2 < \widehat{g}(\widehat{k}^*) < \frac{\widehat{\sigma}^2}{n}, \\ \widehat{\boldsymbol{\theta}}(\cdot, 0), & \text{if } \widehat{g}(\widehat{k}^*) \geq \frac{\widehat{\sigma}^2}{n}. \end{cases} \tag{4.4}$$

4.2 Feasible GRR Estimator

A natural algorithm for choosing the ridge coefficients $k_1, \ldots, k_p$ of the GRR estimator $\widehat{\boldsymbol{\beta}}_p(\mathbf{K}_p)$ is based on the optimum estimator $\widehat{\boldsymbol{\beta}}_p(\mathbf{K}_p^*)$, where $\mathbf{K}_p^* = \text{diag}(k_1^*, \ldots, k_p^*)$ and $k_i^* := \sigma^2/\alpha_i^2$. (See (3.45) and (3.46).) That is,

$$\widehat{k}_i^* := \frac{\widehat{\sigma}^2}{\widehat{\alpha}_i^2}, \quad i = 1, \ldots, p. \tag{4.5}$$

We have the following feasible GRR estimator by plugging $\widehat{k}_i^*$ into the GRR estimator $\widehat{\boldsymbol{\beta}}_p(\mathbf{K}_p)$:

$$\widehat{\boldsymbol{\beta}}_p(\widehat{\mathbf{K}}_p^*) := (\mathbf{X}_p' \mathbf{X}_p + \boldsymbol{\Gamma}_p \widehat{\mathbf{K}}_p^* \boldsymbol{\Gamma}_p')^{-1} \mathbf{X}_p' \mathbf{Y}, \tag{4.6}$$

where $\widehat{\mathbf{K}}_p^* = \text{diag}(\widehat{k}_1^*, \ldots, \widehat{k}_p^*)$.

An algorithm for the ridge coefficient k_0 of the GRR-type estimator $\widehat{\boldsymbol{\theta}}(k_0, \widehat{\mathbf{K}}_p^*)$ is obtained by using (3.52) as follows:

$$\widehat{k}_0^* = \begin{cases} \text{any value greater than } 0, & \text{if } \widehat{g}(\widehat{\mathbf{K}}_p^*) \leq -\widehat{\alpha}_0^2, \\ \widehat{k}_0^*(\widehat{\mathbf{K}}_p^*), & \text{if } -\widehat{\alpha}_0^2 < \widehat{g}(\widehat{\mathbf{K}}_p^*) < \frac{\widehat{\sigma}^2}{n}, \\ 0, & \text{if } \widehat{g}(\widehat{\mathbf{K}}_p^*) \geq \frac{\widehat{\sigma}^2}{n}, \end{cases} \tag{4.7}$$

where

$$\widehat{g}(\widehat{\mathbf{K}}_p^*) := \widehat{\alpha}_0 \boldsymbol{m}_p' \mathbf{S}_p^{-1} \boldsymbol{\Gamma}_p (\boldsymbol{\Lambda}_p(\widehat{\mathbf{K}}_p^*)^{-1} \boldsymbol{\Lambda}_p - \mathbf{I}_p) \widehat{\boldsymbol{\alpha}}_p, \quad \widehat{k}_0^*(\widehat{\mathbf{K}}_p^*) := \frac{n}{\widehat{\alpha}_0^2 + \widehat{g}(\widehat{\mathbf{K}}_p^*)} \left(\frac{\widehat{\sigma}^2}{n} - g(\widehat{\mathbf{K}}_p^*) \right)$$

We obtain the following feasible GRR-type estimator by plugging $\widehat{k}_0$ into the

GRR-type estimator $\widehat{\boldsymbol{\theta}}(k_0, \widehat{\mathbf{K}}_p^*)$:

$$\widehat{\boldsymbol{\theta}}(\widehat{k}_0^*, \widehat{\mathbf{K}}_p^*) = \begin{cases} \widehat{\boldsymbol{\theta}}(k_0, \widehat{\mathbf{K}}_p^*), (k_0 > 0) & \text{if } \widehat{g}(\widehat{\mathbf{K}}_p^*) \leq -\widehat{\alpha}_0^2, \\ \widehat{\boldsymbol{\theta}}(\widehat{k}_0^*(\widehat{\mathbf{K}}_p^*), \widehat{\mathbf{K}}_p^*), & \text{if } -\widehat{\alpha}_0^2 < \widehat{g}(\widehat{\mathbf{K}}_p^*) < \frac{\widehat{\sigma}^2}{n}, \\ \widehat{\boldsymbol{\theta}}(0, \widehat{\mathbf{K}}_p^*), & \text{if } \widehat{g}(\widehat{\mathbf{K}}_p^*) \geq \frac{\widehat{\sigma}^2}{n}. \end{cases} \quad (4.8)$$

4.3 Feasible PCR Estimator

A natural algorithm for choosing the number of principal components r is based on the method embodied in (3.39). That is,

$$\widehat{r}^* := \min\left\{ r \mid \widehat{\tau}_i^2 < 1, \ \forall i \in N_{p\setminus r} \right\}, \quad (4.9)$$

where

$$\widehat{\tau}_i^2 := \frac{\widehat{\alpha}_i^2}{\widehat{\sigma}^2/\lambda_i}.$$

We have the following feasible PCR estimator by plugging $\widehat{r}^*$ into the PCR estimator $\widehat{\boldsymbol{\beta}}_p(r, \cdot)$:

$$\widehat{\boldsymbol{\beta}}_p(\widehat{r}^*, \cdot) := \boldsymbol{\Gamma}_{\widehat{r}^*} \boldsymbol{\Lambda}_{\widehat{r}^*}^{-1} \boldsymbol{\Gamma}_{\widehat{r}^*}' \mathbf{X}_p' \mathbf{Y}. \quad (4.10)$$

The feasible PCR-type estimator for $\boldsymbol{\theta}$ is defined as follows:

$$\widehat{\boldsymbol{\theta}}(\widehat{r}^*, \cdot) := \mathbf{T}^{-1} \widehat{\boldsymbol{\beta}}(\widehat{r}^*, \cdot) = \begin{bmatrix} \widehat{\beta}_0 - \mathbf{m}_p' \mathbf{S}_p^{-1} \widehat{\boldsymbol{\beta}}_p(\widehat{r}^*, \cdot) \\ \mathbf{S}_p^{-1} \widehat{\boldsymbol{\beta}}_p(\widehat{r}^*, \cdot) \end{bmatrix}. \quad (4.11)$$

4.4 Feasible r-k Class Estimator

We consider an algorithm for choosing (r, k) for the r-k class estimator which first chooses r and then k. The sub-algorithm for r is $\widehat{r}^*$ in (4.9), and a sub-algorithm for choosing k is implemented by evaluating

$$\left| \frac{\widehat{\rho}(\widehat{r}^*, k)'}{2} \right| = \left| \sum_{i=1}^{\widehat{r}^*} \frac{\lambda_i(\widehat{\alpha}_i(k)^2 k - \widehat{\sigma}^2)}{(\lambda_i + k)^3} \right| \quad (4.12)$$

for a mesh of k values and selecting that value of k (say $\widehat{k}^*$) associated with the observed minimum. Note that this algorithm is similar to the algorithm for k of the ORR estimator given as (4.1). We have the following feasible r-k class estimator by plugging $\widehat{r}^*$ and $\widehat{k}^*$ into the r-k class estimator $\widehat{\boldsymbol{\beta}}_p(r, k)$:

$$\widehat{\boldsymbol{\beta}}_p(\widehat{r}^*, \widehat{k}^*) := \boldsymbol{\Gamma}_{\widehat{r}^*} \boldsymbol{\Lambda}_{\widehat{r}^*}^{-1}(\widehat{k}^*) \boldsymbol{\Gamma}_{\widehat{r}^*}' \mathbf{X}_p' \mathbf{Y}. \quad (4.13)$$

An algorithm for the ridge coefficient k_0 of the r-k-class-type estimator $\widehat{\boldsymbol{\theta}}(r, k_0)$ is obtained by using (3.53) as follows:

$$\widehat{k}_0^* = \begin{cases} \text{any value greater than } \widehat{k}^*, & \text{if } g(\widehat{r}^*, \widehat{k}^*) \leq -\widehat{\alpha}_0^2, \\ \widehat{k}_0^*(\widehat{r}^*, \widehat{k}^*), & \text{if } -\widehat{\alpha}_0^2 < \widehat{g}(\widehat{r}^*, \widehat{k}^*) < \dfrac{\widehat{\sigma}^2}{n}, \\ 0, & \text{if } \widehat{g}(\widehat{r}^*, \widehat{k}^*) \geq \dfrac{\widehat{\sigma}^2}{n}, \end{cases} \qquad (4.14)$$

where

$$\widehat{g}(\widehat{r}^*, \widehat{k}^*) := \widehat{\alpha}_0 \boldsymbol{m}_p' \mathbf{S}_p^{-1} (\boldsymbol{\Gamma}_{\widehat{r}^*} (\boldsymbol{\Lambda}_{\widehat{r}^*}(\widehat{k}^*))^{-1} \boldsymbol{\Lambda}_{\widehat{r}^*} - \mathbf{I}_{\widehat{r}^*}) \widehat{\boldsymbol{\alpha}}_{\widehat{r}^*} - \boldsymbol{\Gamma}_{p \setminus \widehat{r}^*} \widehat{\boldsymbol{\alpha}}_{p \setminus \widehat{r}^*}),$$

$$\widehat{k}_0^*(\widehat{r}^*, \widehat{k}^*) := \dfrac{n}{\widehat{\alpha}_0^2 + \widehat{g}(\widehat{r}^*, \widehat{k}^*)} \left(\dfrac{\widehat{\sigma}^2}{n} - \widehat{g}(\widehat{r}^*, \widehat{k}^*) \right).$$

We have the following feasible r-k-class-type estimator by plugging $\widehat{k}_0^*$ into the r-k-class-type estimator $\widehat{\boldsymbol{\theta}}(\widehat{r}^*, k_0)$:

$$\widehat{\boldsymbol{\theta}}(\widehat{r}^*, \widehat{k}_0^*) = \begin{cases} \widehat{\boldsymbol{\theta}}(\widehat{r}^*, \widehat{k}_0^*), (\widehat{k}_0^* > \widehat{k}^*) & \text{if } \widehat{g}(\widehat{r}^*, \widehat{k}^*) \leq -\widehat{\alpha}_0^2, \\ \widehat{\boldsymbol{\theta}}(\widehat{r}^*, \widehat{k}_0^*(\widehat{k}^*)), & \text{if } -\widehat{\alpha}_0^2 < \widehat{g}(\widehat{r}^*, \widehat{k}^*) < \dfrac{\widehat{\sigma}^2}{n}, \\ \widehat{\boldsymbol{\theta}}(\widehat{r}^*, 0), & \text{if } \widehat{g}(\widehat{r}^*, \widehat{k}^*) \geq \dfrac{\widehat{\sigma}^2}{n}. \end{cases} \qquad (4.15)$$

Remark 4.1. *It is difficult to obtain the exact moments of the feasible shrinkage regression estimators because the algorithms for choosing r and k, $k_0, k_1, \ldots, k_p$ depend on some random variables. It is also difficult to measure the improvements of the TMSE of the feasible shrinkage regression estimators. In the next chapter, we consider the exact moments of the feasible GRR estimator.*

Chapter 5
Exact Moments of Feasible GRR Estimator

It is difficult to obtain the exact moments of the "feasible" shrinkage regression estimators because the algorithms for choosing the number of principal components r or the ridge coefficients $k, k_0, k_1, \ldots, k_p$ depend on some random variables. There has been little previous work reported on such problems, but Dwivedi et al. (1980), Srivastava and Chaturvedi (1983), and Hemmerle and Carey (1983) are precious exceptions which stand as a challenge to other researchers[1].

In this chapter, we precisely examine the exact moments and several MSE criteria of the feasible GRR estimator under the canonical form (M_A) in Section 1.3. Suppose that the errors ε_i ($i = 1, \ldots, n$) are independent and identically distributed according to the normal distribution $N(0, \sigma^2)$. Note that the setup is the Gauss-Markov one with normality (*normal linear model*) $\text{GMN}(\boldsymbol{Y}, \mathbf{A}\boldsymbol{\alpha}, \sigma^2 \mathbf{I}_n)$. (See Section 1.1.)

5.1 Boundedness of Moments

Let us consider boundedness of the first, second, and cross moments of the feasible GRR estimator. Let $\widehat{\boldsymbol{\alpha}}_p^* := \widehat{\boldsymbol{\alpha}}_p(\widehat{\mathbf{K}}_p^*)$, that is,

$$\widehat{\boldsymbol{\alpha}}_p^* := \begin{bmatrix} \widehat{\alpha}_1^* \\ \vdots \\ \widehat{\alpha}_p^* \end{bmatrix} := \begin{bmatrix} \widehat{\alpha}_1(\widehat{k}_1^*) \\ \vdots \\ \widehat{\alpha}_p(\widehat{k}_p^*) \end{bmatrix} = \widehat{\boldsymbol{\alpha}}_p(\widehat{\mathbf{K}}_p^*) = \left(\boldsymbol{\Lambda}_p + \widehat{\mathbf{K}}_p^*\right)^{-1} \mathbf{A}'\boldsymbol{Y} = \begin{bmatrix} \dfrac{\mathbf{a}_1'\boldsymbol{Y}}{\lambda_1 + \widehat{k}_1^*} \\ \vdots \\ \dfrac{\mathbf{a}_p'\boldsymbol{Y}}{\lambda_p + \widehat{k}_p^*} \end{bmatrix}.$$
(5.1)

It is easy to check that the feasible GRR estimator $\widehat{\alpha}_i^*$ is a shrinkage of the OLS estimator $\widehat{\alpha}_i$ because

$$\widehat{\boldsymbol{\alpha}}_p^* = \left(\boldsymbol{\Lambda}_p + \widehat{\mathbf{K}}_p^*\right)^{-1} \boldsymbol{\Lambda}_p \boldsymbol{\Lambda}_p^{-1} \mathbf{A}_p' \boldsymbol{Y} = \left(\mathbf{I}_p + \boldsymbol{\Lambda}_p^{-1} \widehat{\mathbf{K}}_p^*\right)^{-1} \widehat{\boldsymbol{\alpha}}_p$$

$$\Longleftrightarrow \widehat{\alpha}_i^* = \frac{\lambda_i}{\lambda_i + \widehat{k}_i^*} \widehat{\alpha}_i, \quad \left(0 < \frac{\lambda_i}{\lambda_i + \widehat{k}_i^*} < 1\right) \quad i = 1, \ldots, p,$$

---

[1] Ohtani (1993) and Inoue (1999) considered the marginal density function of the feasible GRR estimators.

where
$$\widehat{\boldsymbol{\alpha}}_p \sim N_p(\boldsymbol{\alpha}_p, \sigma^2 \boldsymbol{\Lambda}_p^{-1}) \iff \widehat{\alpha}_i \stackrel{\text{ind.}}{\sim} N\left(\alpha_i, \frac{\sigma^2}{\lambda_i}\right), i = 1, \ldots, p,$$

by $\varepsilon \sim N_n(\mathbf{0}, \sigma^2 \mathbf{I}_n)$, and "$X_i \stackrel{\text{ind.}}{\sim} F_i$" ($i = 1, \ldots, n$) denotes that the X_i's are independently distributed as F_i.

By defining the random variables

$$Z_i := \frac{\widehat{\alpha}_i}{\sigma_{\widehat{\alpha}_i}} \stackrel{\text{ind.}}{\sim} N(\tau_i, 1), \ i = 1, \ldots, p, \quad V := \frac{\nu \widehat{\sigma}^2}{\sigma^2} \sim \chi_\nu^2, \quad (Z_i \perp\!\!\!\perp V; \text{ independent}),$$

we can represent the feasible GRR estimator as follows:

$$\widehat{\alpha}_i^* = \sigma_{\widehat{\alpha}_i} \frac{Z_i^2}{Z_i^2 + V/\nu} Z_i, \tag{5.2}$$

where $\nu := n - p - 1$ is the degree of freedom and

$$\sigma_{\widehat{\alpha}_i} := \sqrt{\mathrm{V}(\widehat{\alpha}_i)} = \frac{\sigma}{\sqrt{\lambda_i}}; \quad \text{standard error,} \tag{5.3}$$

$$\tau_i := \frac{\alpha_i}{\sigma_{\widehat{\alpha}_i}} = \frac{\alpha_i}{\sigma/\sqrt{\lambda_i}}; \quad \text{non-centrality parameter,} \tag{5.4}$$

and we use

$$\frac{\lambda_i}{\lambda_i + \widehat{k}_i^*} = \frac{\lambda_i}{\lambda_i + \widehat{\sigma}^2/\widehat{\alpha}_i^2} = \frac{(\widehat{\alpha}_i/\sigma_{\widehat{\alpha}_i})^2}{(\widehat{\alpha}_i/\sigma_{\widehat{\alpha}_i})^2 + (\widehat{\sigma}^2/\sigma^2)} = \frac{Z_i^2}{Z_i^2 + V/\nu},$$

$$\widehat{\alpha}_i = \sigma_{\widehat{\alpha}_i} \frac{\widehat{\alpha}_i}{\sigma_{\widehat{\alpha}_i}} = \sigma_{\widehat{\alpha}_i} Z_i.$$

The following proposition and corollary guarantee the existence of the first and second moments of the feasible GRR estimator $\widehat{\alpha}_i^*$ and the cross moment between $\widehat{\alpha}_i^*$ and $\widehat{\alpha}_j^*$ ($i \neq j$).

Proposition 5.1. *The first and second absolute moments of the feasible GRR estimator $\widehat{\alpha}_i^*$ are bounded:*

$$E(|\widehat{\alpha}_i^*|) < \infty, \quad E(|\widehat{\alpha}_i^*|^2) < \infty.$$

Additionally, the cross absolute moment between $\widehat{\alpha}_i^*$ and $\widehat{\alpha}_j^*$ ($i \neq j = 1, \ldots, p$) is also bounded:

$$E(|\widehat{\alpha}_i^* \widehat{\alpha}_j^*|) < \infty.$$

Proof. From $0 \leq Z_i^2/(Z_i^2 + V/\nu) < 1$ almost surely (a.s.) and (5.2),

$$\mathrm{E}(|\widehat{\alpha}_i^*|) = \mathrm{E}\left(\left|\sigma_{\widehat{\alpha}_i} \frac{Z_i^2}{Z_i^2 + V/\nu} Z_i\right|\right) = \sigma_{\widehat{\alpha}_i} \mathrm{E}\left(\left|\frac{Z_i^2}{Z_i^2 + V/\nu}\right| |Z_i|\right) < \sigma_{\widehat{\alpha}_i} \mathrm{E}(|Z_i|)$$
$$= \sigma_{\widehat{\alpha}_i} \mu(\tau_i) < \infty,$$

where
$$\mu(\tau_i) := \mathrm{E}(|Z_i|) = \int_{-\infty}^{\infty} |z_i|\phi(z_i - \tau_i)dz_i$$
$$= 2\phi(\tau_i) + \tau_i(\Phi(\tau_i) - \Phi(-\tau_i)) < \infty, \quad \tau_i \in \mathbb{R},$$

and $\phi(\cdot)$ and $\Phi(\cdot)$ are the probability density function and the cumulative distribution function of the standard normal distribution $N(0,1)$, respectively.

Similarly, we have
$$\mathrm{E}(|\widehat{\alpha}_i^*|^2) = \mathrm{E}\left(\sigma_{\widehat{\alpha}_i}^2 \left|\frac{Z_i^2}{Z_i^2 + V/\nu}\right|^2 |Z_i|^2\right) < \sigma_{\widehat{\alpha}_i}^2 \mathrm{E}(|Z_i|^2)$$
$$= \sigma_{\widehat{\alpha}_i}^2 (1 + \tau_i^2) < \infty$$

by using
$$\mathrm{E}(|Z_i|^2) = \int_{-\infty}^{\infty} |z_i|^2 \phi(z_i - \tau_i) dz_i = 1 + \tau_i^2 < \infty.$$

We also know
$$\mathrm{E}(|\widehat{\alpha}_i^* \widehat{\alpha}_j^*|) = \mathrm{E}\left(\sigma_{\widehat{\alpha}_i} \sigma_{\widehat{\alpha}_j} \left|\frac{Z_i^2}{Z_i^2 + V/\nu}\right| \left|\frac{Z_j^2}{Z_j^2 + V/\nu}\right| |Z_i||Z_j|\right) < \sigma_{\widehat{\alpha}_i} \sigma_{\widehat{\alpha}_j} \mathrm{E}(|Z_i||Z_j|)$$
$$= \sigma_{\widehat{\alpha}_i} \sigma_{\widehat{\alpha}_j} \mu(\tau_i) \mu(\tau_j) < \infty.$$

The proof is complete. $\square$

The following corollary is easily obtained from Proposition 5.1:

Corollary 5.1. *The first and second moments of the feasible GRR estimator $\widehat{\alpha}_i^*$ and the cross moment between $\widehat{\alpha}_i^*$ and $\widehat{\alpha}_j^*$ are bounded:*
$$|E(\widehat{\alpha}_i^*)| < \infty, \quad |E(\widehat{\alpha}_i^{*2})| < \infty, \quad |E(\widehat{\alpha}_i^* \widehat{\alpha}_j^*)| < \infty.$$

5.2 First and Second Moments

We know the existence of the moments of the feasible GRR estimator from Corollary 5.1. However, if we would like to calculate them in practice, then we have to consider them separately. Dwivedi *et al.* (1980) gave the first and second moments by using the following idea.

First of all, we *formally* expand the feasible GRR estimator $\widehat{\alpha}_i^*$ to the following series:
$$\widehat{\alpha}_i^* = \sigma_{\widehat{\alpha}_i} \frac{Z_i^3}{Z_i^2 + V/\nu} = \sigma_{\widehat{\alpha}_i} \frac{Z_i^3}{Z_i^2 + V}\left(1 - \frac{\nu - 1}{\nu} \frac{V}{Z_i^2 + V}\right)^{-1} \quad (5.5)$$
$$= \sigma_{\widehat{\alpha}_i} \frac{Z_i^3}{Z_i^2 + V} \sum_{\ell=0}^{\infty} \left(\frac{\nu-1}{\nu}\right)^\ell \left(\frac{V}{Z_i^2 + V}\right)^\ell \quad (5.6)$$
$$= \sigma_{\widehat{\alpha}_i} \sum_{\ell=0}^{\infty} \left(\frac{\nu-1}{\nu}\right)^\ell \frac{Z_i^3 V^\ell}{(Z_i^2 + V)^{\ell+1}},$$

where the following fact is used:

$$|r| < 1 \implies \sum_{\ell=0}^{\infty} r^{\ell} = \frac{1}{1-r}.$$

Similarly, we also expand $\widehat{\alpha}_i^{*2}$ to a series, as follows:

$$\widehat{\alpha}_i^{*2} = \sigma_{\widehat{\alpha}_i}^2 \frac{Z_i^6}{(Z_i^2 + V/\nu)^2} = \sigma_{\widehat{\alpha}_i}^2 \frac{Z_i^6}{(Z_i^2 + V)^2} \left(1 - \frac{\nu-1}{\nu} \frac{V}{Z_i^2 + V}\right)^{-2} \quad (5.7)$$

$$= \sigma_{\widehat{\alpha}_i}^2 \frac{Z_i^6}{(Z_i^2 + V)^2} \sum_{\ell=0}^{\infty} (\ell+1) \left(\frac{\nu-1}{\nu}\right)^{\ell} \left(\frac{V}{Z_i^2 + V}\right)^{\ell} \quad (5.8)$$

$$= \sigma_{\widehat{\alpha}_i}^2 \sum_{\ell=0}^{\infty} (\ell+1) \left(\frac{\nu-1}{\nu}\right)^{\ell} \frac{Z_i^6 V^{\ell}}{(Z_i^2 + V)^{\ell+2}},$$

where we use that

$$|r| < 1 \implies \sum_{\ell=0}^{\infty} (\ell+1) r^{\ell} = \frac{1}{(1-r)^2}.$$

Next, we take the expectations of these estimators. If exchangeability between summation and expectation holds, then we have

$$\mathrm{E}(\widehat{\alpha}_i^*) = \sigma_{\widehat{\alpha}_i} \sum_{\ell=0}^{\infty} \left(\frac{\nu-1}{\nu}\right)^{\ell} \mathrm{E}\left(\frac{Z_i^3 V^{\ell}}{(Z_i^2 + V)^{\ell+1}}\right), \quad (5.9)$$

$$\mathrm{E}(\widehat{\alpha}_i^{*2}) = \sigma_{\widehat{\alpha}_i}^2 \sum_{\ell=0}^{\infty} (\ell+1) \left(\frac{\nu-1}{\nu}\right)^{\ell} \mathrm{E}\left(\frac{Z_i^6 V^{\ell}}{(Z_i^2 + V)^{\ell+2}}\right). \quad (5.10)$$

Finally, by using Lemma B.1 in Appendix B, we know the moments as follows:

$$\mathrm{E}(\widehat{\alpha}_i^*) = \sigma_{\widehat{\alpha}_i} \tau_i \sum_{\ell=0}^{\infty} \left(\frac{\nu-1}{\nu}\right)^{\ell} \sum_{m=0}^{\infty} \frac{B\left(\ell + \frac{\nu}{2}, m + \frac{5}{2}\right)}{B\left(\frac{\nu}{2}, m + \frac{3}{2}\right)} p(m; \tau_i^2/2)$$

$$= \alpha_i \sum_{\ell=0}^{\infty} \left(\frac{\nu-1}{\nu}\right)^{\ell} \sum_{m=0}^{\infty} \frac{B\left(\ell + \frac{\nu}{2}, m + \frac{5}{2}\right)}{B\left(\frac{\nu}{2}, m + \frac{3}{2}\right)} p(m; \delta_i), \quad (5.11)$$

$$\mathrm{E}(\widehat{\alpha}_i^{*2}) = \sigma_{\widehat{\alpha}_i}^2 \sum_{\ell=0}^{\infty} (\ell+1) \left(\frac{\nu-1}{\nu}\right)^{\ell} \sum_{m=0}^{\infty} (2m+1) \frac{B\left(\ell + \frac{\nu}{2}, m + \frac{7}{2}\right)}{B\left(\frac{\nu}{2}, m + \frac{3}{2}\right)} p(m; \tau_i^2/2)$$

$$= \alpha_i^2 \frac{1}{2\delta_i} \sum_{\ell=0}^{\infty} (\ell+1) \left(\frac{\nu-1}{\nu}\right)^{\ell} \sum_{m=0}^{\infty} (2m+1) \frac{B\left(\ell + \frac{\nu}{2}, m + \frac{7}{2}\right)}{B\left(\frac{\nu}{2}, m + \frac{3}{2}\right)} p(m; \delta_i),$$

$$(5.12)$$

where
$$\delta_i := \frac{\tau_i^2}{2} = \frac{1}{2}\frac{\alpha_i^2}{\sigma^2/\lambda_i}, \qquad (5.13)$$

$B(\cdot,\cdot)$ is the beta function defined by
$$B(x,y) := \int_0^1 t^{x-1}(1-t)^{y-1}dt,$$

and $p(m;\delta_i)$ is the probability mass function of the Poisson distribution $P_o(\delta_i)$ defined by
$$p(m;\delta_i) := \frac{\delta_i^m}{m!}e^{-\delta_i}.$$

Recall that
$$\sigma_{\widehat{\alpha}_i} = \sqrt{V(\widehat{\alpha}_i)} = \frac{\sigma}{\sqrt{\lambda_i}}, \quad \tau_i = \frac{\alpha_i}{\sigma_{\widehat{\alpha}_i}} = \frac{\alpha_i}{\sigma/\sqrt{\lambda_i}}.$$

Note that the forms in (5.11), (5.12) are essentially equivalent to the results given by Dwivedi, Srivastava and Hall (1980).

For the above process, we have the following questions:

(Q1) Are summation and expectation in (5.9) and (5.10) exchangeable?
(Q2) Do the double series in (5.11) and (5.12) converge?

To answer question (Q1), we first prepare some lemmas. Let us define the following:

$$\widehat{\alpha}_{iL}^* := \sigma_{\widehat{\alpha}_i}\frac{Z_i^3}{Z_i^2+V}\sum_{\ell=0}^{L}\left(\frac{\nu-1}{\nu}\right)^\ell\left(\frac{V}{Z_i^2+V}\right)^\ell, \qquad (5.14)$$

$$\widehat{\alpha}_{iL}^{*2} := \sigma_{\widehat{\alpha}_i}^2\frac{Z_i^6}{(Z_i^2+V)^2}\sum_{\ell=0}^{L}(\ell+1)\left(\frac{\nu-1}{\nu}\right)^\ell\left(\frac{V}{Z_i^2+V}\right)^\ell, \qquad (5.15)$$

where $\widehat{\alpha}_{iL}^{*2} \neq (\widehat{\alpha}_{iL}^*)^2$. Note that we can write

$$\widehat{\alpha}_{iL}^* = \widehat{\alpha}_i^* - \sigma_{\widehat{\alpha}_i}Z_i\frac{Z_i^2}{Z_i^2+V}\frac{\left(\frac{\nu-1}{\nu}\right)^{L+1}\left(\frac{V}{Z_i^2+V}\right)^{L+1}}{1-\left(\frac{\nu-1}{\nu}\right)\left(\frac{V}{Z_i^2+V}\right)}, \qquad (5.16)$$

$$\widehat{\alpha}_{iL}^{*2} = \widehat{\alpha}_i^{*2} - \sigma_{\widehat{\alpha}_i}^2 Z_i^2\left(\frac{Z_i^2}{Z_i^2+V}\right)^2$$
$$\times \frac{\left(\frac{\nu-1}{\nu}\right)^{L+1}\left(\frac{V}{Z_i^2+V}\right)^{L+1}\left\{(L+1)\left(1-\left(\frac{\nu-1}{\nu}\right)\left(\frac{V}{Z_i^2+V}\right)\right)+1\right\}}{\left\{1-\left(\frac{\nu-1}{\nu}\right)\left(\frac{V}{Z_i^2+V}\right)\right\}^2}, \qquad (5.17)$$

by using (5.5), (5.7), and

$$\sum_{\ell=0}^{L}r^\ell = \frac{1}{1-r} - \frac{r^{L+1}}{1-r}, \quad \sum_{\ell=0}^{L}(\ell+1)r^\ell = \frac{1}{(1-r)^2} - \frac{r^{L+1}\{(L+1)(1-r)+1\}}{(1-r)^2}.$$

Lemma 5.1. *Let $\nu \geq 2$. For any $\varepsilon > 0$,*

$$\sum_{L=0}^{\infty} P(B_{1L}(\varepsilon)) < \infty, \quad \sum_{L=0}^{\infty} P(B_{2L}(\varepsilon)) < \infty,$$

where

$$B_{1L}(\varepsilon) := \{|\widehat{\alpha}_{iL}^* - \widehat{\alpha}_i^*| > \varepsilon\}, \quad B_{2L}(\varepsilon) := \{|\widehat{\alpha}_{iL}^{*2} - \widehat{\alpha}_i^{*2}| > \varepsilon\}.$$

Proof. Note that for any $\nu \geq 2$,

$$0 \leq \frac{Z_i^2}{Z_i^2 + V} < 1, \quad 0 < \frac{V}{Z_i^2 + V} \leq 1, \quad 0 < \frac{\nu-1}{\nu} \frac{V}{Z_i^2 + V} \leq \frac{\nu-1}{\nu} (<1).$$

By using these results and (5.16),

$$P(|\widehat{\alpha}_{iL}^* - \widehat{\alpha}_i^*| > \varepsilon) = P\left(\sigma_{\widehat{\alpha}_i} |Z_i| \frac{Z_i^2}{Z_i^2 + V} \frac{\left(\frac{\nu-1}{\nu}\right)^{L+1} \left(\frac{V}{Z_i^2+V}\right)^{L+1}}{1 - \left(\frac{\nu-1}{\nu}\right)\left(\frac{V}{Z_i^2+V}\right)} > \varepsilon\right)$$

$$< P\left(\sigma_{\widehat{\alpha}_i} |Z_i| \frac{\left(\frac{\nu-1}{\nu}\right)^{L+1} \left(\frac{V}{Z_i^2+V}\right)^{L+1}}{1 - \left(\frac{\nu-1}{\nu}\right)\left(\frac{V}{Z_i^2+V}\right)} > \varepsilon\right)$$

$$= P\left(\sigma_{\widehat{\alpha}_i} |Z_i| \left(\frac{\nu-1}{\nu}\right)^{L+1} \left(\frac{V}{Z_i^2+V}\right)^{L+1} > \left(1 - \frac{\nu-1}{\nu} \frac{V}{Z_i^2+V}\right)\varepsilon\right)$$

$$\leq P\left(\sigma_{\widehat{\alpha}_i} |Z_i| \left(\frac{\nu-1}{\nu}\right)^{L+1} > \frac{\varepsilon}{\nu}\right)$$

$$= P\left(|Z_i| > \frac{\varepsilon}{\nu \sigma_{\widehat{\alpha}_i}} \left(\frac{\nu}{\nu-1}\right)^{L+1}\right),$$

and we know

$$P(B_{1L}(\varepsilon)) < P\left(|Z_i| > \frac{\varepsilon}{\nu \sigma_{\widehat{\alpha}_i}} \left(\frac{\nu}{\nu-1}\right)^{L+1}\right) \leq \frac{\nu \sigma_{\widehat{\alpha}_i}}{\varepsilon} \left(\frac{\nu-1}{\nu}\right)^{L+1} E(|Z_i|) =: a_{1L}$$

by *Markov's inequality*. Hence,

$$\sum_{L=0}^{\infty} P(B_{1L}(\varepsilon)) < \sum_{L=0}^{\infty} a_{1L} = \sum_{L=0}^{\infty} \frac{\nu \sigma_{\widehat{\alpha}_i}}{\varepsilon} \left(\frac{\nu-1}{\nu}\right)^{L+1} \mu(\tau_i) = \frac{\sigma_{\widehat{\alpha}_i} \mu(\tau_i)}{\varepsilon} \nu(\nu-1) < \infty,$$

where $\mu(\tau_i) := E(|Z_i|)$ and we use that

$$\sum_{L=0}^{\infty} \left(\frac{\nu-1}{\nu}\right)^L = \frac{1}{1 - \left(\frac{\nu-1}{\nu}\right)} = \nu.$$

Similarly, we know
$\mathrm{P}\left(|\widehat{\alpha}_{iL}^{*2} - \widehat{\alpha}_i^{*2}| > \varepsilon\right)$

$$= \mathrm{P}\left(\sigma_{\widehat{\alpha}_i}^2 Z_i^2 \left(\frac{Z_i^2}{Z_i^2+V}\right)^2 \frac{\left(\frac{\nu-1}{\nu}\right)^{L+1}\left(\frac{V}{Z_i^2+V}\right)^{L+1}\left\{(L+1)\left(1-\left(\frac{\nu-1}{\nu}\right)\left(\frac{V}{Z_i^2+V}\right)\right)+1\right\}}{\left\{1-\left(\frac{\nu-1}{\nu}\right)\left(\frac{V}{Z_i^2+V}\right)\right\}^2} > \varepsilon\right)$$

$$< \mathrm{P}\left(\sigma_{\widehat{\alpha}_i}^2 Z_i^2 \left(\frac{\nu-1}{\nu}\right)^{L+1}(L+2) > \left\{1-\left(\frac{\nu-1}{\nu}\right)\right\}^2 \varepsilon\right)$$

$$= \mathrm{P}\left(Z_i^2 > \frac{\varepsilon}{\nu^2 \sigma_{\widehat{\alpha}_i}^2 (L+2)}\left(\frac{\nu}{\nu-1}\right)^{L+1}\right)$$

and by Markov's inequality,

$$\mathrm{P}(B_{2L}(\varepsilon)) < \mathrm{P}\left(Z_i^2 > \frac{\varepsilon}{\nu^2 \sigma_{\widehat{\alpha}_i}^2 (L+2)}\left(\frac{\nu}{\nu-1}\right)^{L+1}\right)$$

$$\leq \frac{\nu^2 \sigma_{\widehat{\alpha}_i}^2 (L+2)}{\varepsilon}\left(\frac{\nu-1}{\nu}\right)^{L+1} \mathrm{E}(Z_i^2)$$

$$= \frac{\nu^2 \sigma_{\widehat{\alpha}_i}^2 (1+\tau_i^2)}{\varepsilon}(L+2)\left(\frac{\nu-1}{\nu}\right)^{L+1} =: a_{2L}.$$

Hence,

$$\sum_{L=0}^{\infty} \mathrm{P}(B_{2L}(\varepsilon)) < \sum_{L=0}^{\infty} a_{2L} = \frac{\sigma_{\widehat{\alpha}_i}^2 (1+\tau_i^2)}{\varepsilon}\nu^2(\nu^2-1) < \infty,$$

where the following fact is used:

$$\sum_{L=0}^{\infty}(L+2)\left(\frac{\nu-1}{\nu}\right)^{L+1} = \nu^2 - 1.$$

The proof is complete. □

Lemma 5.2. *Let* $\nu \geq 2$. $\widehat{\alpha}_{iL}^*$ *and* $\widehat{\alpha}_{iL}^{*2}$ *almost surely converge to* $\widehat{\alpha}_i^*$ *and* $\widehat{\alpha}_i^{*2}$, *respectively:*

$$\lim_{L \to \infty} \widehat{\alpha}_{iL}^* = \widehat{\alpha}_i^* \ a.s., \quad \lim_{L \to \infty} \widehat{\alpha}_{iL}^{*2} = \widehat{\alpha}_i^{*2} \ a.s.$$

Proof. By using Lemma 5.1 and *Borel-Cantelli's lemma*, we know

$$\sum_{L=0}^{\infty} \mathrm{P}(B_{1L}(\varepsilon)) < \infty \Rightarrow \mathrm{P}(\overline{\lim_{L \to \infty}} B_{1L}(\varepsilon)) = 0$$

for any $\varepsilon > 0$, where

$$\overline{\lim_{L \to \infty}} B_{1L}(\varepsilon) := \bigcap_{k=1}^{\infty} \bigcup_{L=k}^{\infty} B_{1L}(\varepsilon).$$

Hence, $\widehat{\alpha}^*_{iL}$ almost surely converge to $\widehat{\alpha}^*_i$. Almost sure convergence of $\widehat{\alpha}^{*2}_{iL}$ is also proved. □

Lemma 5.3. *There exist some random variables W_1 and W_2 such that $E(W_1) < \infty$, $E(W_2) < \infty$ and $|\widehat{\alpha}^*_{iL}| < W_1$ a.s., $|\widehat{\alpha}^{*2}_{iL}| < W_2$ a.s. for any $L \geq 0$.*

Proof. Let us set $W_1 := \sigma_{\widehat{\alpha}_i} \nu |Z_i|$. Then,

$$E(W_1) = \sigma_{\widehat{\alpha}_i} \nu \int_{-\infty}^{\infty} |z_i| \phi(z_i - \tau_i) dz_i < \infty,$$

and

$$|\widehat{\alpha}^*_{iL}| \leq \sigma_{\widehat{\alpha}_i} |Z_i| \frac{Z_i^2}{Z_i^2 + V} \sum_{\ell=0}^{L} \left(\frac{\nu-1}{\nu}\right)^\ell \left(\frac{V}{Z_i^2 + V}\right)^\ell$$

$$< \sigma_{\widehat{\alpha}_i} |Z_i| \sum_{\ell=0}^{\infty} \left(\frac{\nu-1}{\nu}\right)^\ell = W_1 \quad \text{a.s.}$$

Similarly, if we set $W_2 := \sigma^2_{\widehat{\alpha}_i} \nu^2 Z_i^2$,

$$E(W_2) = \sigma^2_{\widehat{\alpha}_i} \nu^2 \int_{-\infty}^{\infty} z_i^2 \phi(z_i - \tau_i) dz_i = \sigma^2_{\widehat{\alpha}_i} \nu^2 (1 + \tau_i^2) < \infty,$$

and

$$|\widehat{\alpha}^{*2}_{iL}| \leq \sigma^2_{\widehat{\alpha}_i} Z_i^2 \left(\frac{Z_i^2}{Z_i^2 + V}\right)^2 \sum_{\ell=0}^{L} (\ell+1) \left(\frac{\nu-1}{\nu}\right)^\ell \left(\frac{V}{Z_i^2 + V}\right)^\ell$$

$$< \sigma^2_{\widehat{\alpha}_i} Z_i^2 \sum_{\ell=0}^{\infty} (\ell+1) \left(\frac{\nu-1}{\nu}\right)^\ell = W_2 \quad \text{a.s.}$$

The proof is complete. □

Question (Q1) will be answered by Theorem 5.1 as follows:

Theorem 5.1. *Let $\nu \geq 2$. Then,*

$$E(\widehat{\alpha}^*_i) = \lim_{L \to \infty} E(\widehat{\alpha}^*_{iL}) = \lim_{L \to \infty} \sigma_{\widehat{\alpha}_i} \sum_{\ell=0}^{L} \left(\frac{\nu-1}{\nu}\right)^\ell E\left(\frac{Z_i^3 V^\ell}{(Z_i^2 + V)^{\ell+1}}\right), \quad (5.18)$$

$$E(\widehat{\alpha}^{*2}_i) = \lim_{L \to \infty} E(\widehat{\alpha}^{*2}_{iL}) = \lim_{L \to \infty} \sigma^2_{\widehat{\alpha}_i} \sum_{\ell=0}^{L} (\ell+1) \left(\frac{\nu-1}{\nu}\right)^\ell E\left(\frac{Z_i^6 V^\ell}{(Z_i^2 + V)^{\ell+2}}\right). \quad (5.19)$$

Proof. Since we can use *Lebesgue's dominated convergence theorem* (e.g. Billingsley (1995)) by Lemmas 5.2, 5.3, the theorem holds. □

5.2. FIRST AND SECOND MOMENTS

Next, we give the following proposition for answering question (Q2):

Proposition 5.2. *The following series converge for any $\nu \geq 1$, $\delta_i > 0$:*

$$\mu_1(\delta_i, \nu) := \sum_{\ell=0}^{\infty} \left(\frac{\nu-1}{\nu}\right)^\ell \sum_{m=0}^{\infty} \frac{B\left(\ell + \frac{\nu}{2}, m + \frac{5}{2}\right)}{B\left(\frac{\nu}{2}, m + \frac{3}{2}\right)} p(m; \delta_i), \qquad (5.20)$$

$$\mu_2(\delta_i, \nu) := \frac{1}{2\delta_i} \sum_{\ell=0}^{\infty} (\ell+1) \left(\frac{\nu-1}{\nu}\right)^\ell \sum_{m=0}^{\infty} (2m+1) \frac{B\left(\ell + \frac{\nu}{2}, m + \frac{7}{2}\right)}{B\left(\frac{\nu}{2}, m + \frac{3}{2}\right)} p(m; \delta_i). \qquad (5.21)$$

Proof. Since the beta function is monotonically decreasing for $\ell, m \geq 0$,

$$\frac{B\left(\ell + \frac{\nu}{2}, m + \frac{5}{2}\right)}{B\left(\frac{\nu}{2}, m + \frac{3}{2}\right)} < 1, \quad \frac{B\left(\ell + \frac{\nu}{2}, m + \frac{7}{2}\right)}{B\left(\frac{\nu}{2}, m + \frac{3}{2}\right)} < 1.$$

(See Proposition A.4 in Appendix A.) Hence,

$$\mu_1(\delta_i, \nu) < \sum_{\ell=0}^{\infty} \left(\frac{\nu-1}{\nu}\right)^\ell \sum_{m=0}^{\infty} p(m; \delta_i) = \nu < \infty,$$

$$\mu_2(\delta_i, \nu) < \frac{1}{2\delta_i} \sum_{\ell=0}^{\infty} (\ell+1) \left(\frac{\nu-1}{\nu}\right)^\ell \sum_{m=0}^{\infty} (2m+1) p(m; \delta_i) = \nu^2 \left(1 + \frac{1}{2\delta_i}\right) < \infty.$$

That is, the series are convergent because the series are positive. □

Question (Q2) has then also been answered, by Proposition 5.2.

Remark 5.1. *If $\alpha_i = 0$, then the optimum value $k_i^* = \sigma^2/\alpha_i^2$ for the GRR estimator $\widehat{\alpha}_i^*$ does not exist* [2]. *However, the moments of the feasible GRR estimators $\widehat{\alpha}_i^*$ can be obtained. The first moment is*

$$E(\widehat{\alpha}_i^*) = \alpha_i \mu_{1i}(\delta_i, \nu) = \alpha_i \sum_{\ell=0}^{\infty} \left(\frac{\nu-1}{\nu}\right)^\ell \sum_{m=0}^{\infty} \frac{B\left(\ell + \frac{\nu}{2}, m + \frac{5}{2}\right)}{B\left(\frac{\nu}{2}, m + \frac{3}{2}\right)} p(m; \delta_i)$$

$$= \alpha_i \sum_{\ell=0}^{\infty} \left(\frac{\nu-1}{\nu}\right)^\ell \frac{B\left(\ell + \frac{\nu}{2}, \frac{5}{2}\right)}{B\left(\frac{\nu}{2}, \frac{3}{2}\right)}$$

$$= 0,$$

because

$$\alpha_i = 0 \implies \delta_i = \frac{1}{2} \frac{\alpha_i^2}{\sigma^2/\lambda_i} = 0, \quad p(m; 0) = \begin{cases} 1, & m = 0, \\ 0, & m \neq 0. \end{cases}$$

---

[2] In this case, the derivative of the MSE of the GRR estimator is monotonically decreasing with respect to k_i. Therefore, $k_i^* = \infty$.

Note that $\widehat{\alpha}_i^*$ is an unbiased estimator of $\alpha_i(=0)$.

Similarly, the second moment is

$$E(\widehat{\alpha}_i^{*2}) = \frac{\sigma^2}{\lambda_i} \sum_{\ell=0}^{\infty} (\ell+1) \left(\frac{\nu-1}{\nu}\right)^{\ell} \sum_{m=0}^{\infty} (2m+1) \frac{B\left(\ell+\frac{\nu}{2}, m+\frac{7}{2}\right)}{B\left(\frac{\nu}{2}, m+\frac{3}{2}\right)} p(m; \delta_i)$$

$$= \frac{\sigma^2}{\lambda_i} \sum_{\ell=0}^{\infty} (\ell+1) \left(\frac{\nu-1}{\nu}\right)^{\ell} \frac{B\left(\ell+\frac{\nu}{2}, \frac{7}{2}\right)}{B\left(\frac{\nu}{2}, \frac{3}{2}\right)},$$

where

$$\widetilde{\mu}_2(\nu) := \sum_{\ell=0}^{\infty} (\ell+1) \left(\frac{\nu-1}{\nu}\right)^{\ell} \frac{B\left(\ell+\frac{\nu}{2}, \frac{7}{2}\right)}{B\left(\frac{\nu}{2}, \frac{3}{2}\right)} \tag{5.22}$$

is a monotonically decreasing function of ν. (See Figure 5.1.)

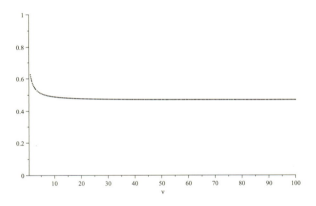

Figure 5.1: Plot of $\widetilde{\mu}_2(\nu)$

5.3 Cross Moments

Let us consider the cross moment between the feasible GRR estimators $\widehat{\alpha}_i^*$ and $\widehat{\alpha}_j^*$ ($i \neq j$). By using (5.6), we can write

$$\widehat{\alpha}_i^* \widehat{\alpha}_j^* = \sigma_{\widehat{\alpha}_i} \sigma_{\widehat{\alpha}_j} \sum_{\ell_1=0}^{\infty} \left(\frac{\nu-1}{\nu}\right)^{\ell_1} \frac{Z_i^3 V^{\ell_1}}{(Z_i^2+V)^{\ell_1+1}} \sum_{\ell_2=0}^{\infty} \left(\frac{\nu-1}{\nu}\right)^{\ell_2} \frac{Z_j^3 V^{\ell_2}}{(Z_j^2+V)^{\ell_2+1}}$$

$$= \sigma_{\widehat{\alpha}_i} \sigma_{\widehat{\alpha}_j} \sum_{\ell_1=0}^{\infty} \sum_{\ell_2=0}^{\infty} \left(\frac{\nu-1}{\nu}\right)^{\ell_1+\ell_2} \frac{Z_i^3 V^{\ell_1}}{(Z_i^2+V)^{\ell_1+1}} \frac{Z_j^3 V^{\ell_2}}{(Z_j^2+V)^{\ell_2+1}}.$$

5.3. CROSS MOMENTS

If we allow an exchange of the summation and expectation, then we have

$$
\mathrm{E}\left(\widehat{\alpha}_i^* \widehat{\alpha}_j^*\right) = \sigma_{\widehat{\alpha}_i} \sigma_{\widehat{\alpha}_j} \sum_{\ell_1=0}^{\infty} \sum_{\ell_2=0}^{\infty} \left(\frac{\nu-1}{\nu}\right)^{\ell_1+\ell_2} \mathrm{E}\left(\frac{Z_i^3 V^{\ell_1}}{(Z_i^2+V)^{\ell_1+1}} \frac{Z_j^3 V^{\ell_2}}{(Z_j^2+V)^{\ell_2+1}}\right). \tag{5.23}
$$

Hence, we obtain the following theorem by Lemma B.2 in Appendix B:

Theorem 5.2. *The cross moment between the feasible GRR estimators $\widehat{\alpha}_i^*$ and $\widehat{\alpha}_j^*$ ($i \neq j$) is given by*

$$
\mathrm{E}\left(\widehat{\alpha}_i^* \widehat{\alpha}_j^*\right) = \alpha_i \alpha_j \sum_{\ell_1=0}^{\infty} \sum_{\ell_2=0}^{\infty} \left(\frac{\nu-1}{\nu}\right)^{\ell_1+\ell_2}
$$
$$
\times \sum_{m_1=0}^{\infty} \sum_{m_2=0}^{\infty} \frac{\mathcal{B}_{\ell_1+1\ \ell_2+1}\left(m_1+\frac{5}{2}, m_2+\frac{5}{2}, \ell_1+\ell_2+\frac{\nu}{2}\right)}{B\left(m_1+\frac{3}{2}, m_2+\frac{3}{2}, \frac{\nu}{2}\right)} p(m_1; \delta_i)\, p(m_2; \delta_j), \tag{5.24}
$$

where

$$
B\left(m_1+\frac{3}{2}, m_2+\frac{3}{2}, \frac{\nu}{2}\right) = \frac{\Gamma\left(m_1+\frac{3}{2}\right)\Gamma\left(m_2+\frac{3}{2}\right)\Gamma\left(\frac{\nu}{2}\right)}{\Gamma\left(m_1+m_2+3+\frac{\nu}{2}\right)}
$$

(see Section A.2 in Appendix A),

$$
\mathcal{B}_{\ell_1+1\ \ell_2+1}\left(m_1+\frac{5}{2}, m_2+\frac{5}{2}, \ell_1+\ell_2+\frac{\nu}{2}\right)
$$
$$
= \iint_{\mathcal{D}} \frac{1}{(1-t_1)^{\ell_2+1}} \frac{1}{(1-t_2)^{\ell_1+1}} t_1^{m_1+\frac{5}{2}-1} t_2^{m_2+\frac{5}{2}-1} (1-t_1-t_2)^{\ell_1+\ell_2+\frac{\nu}{2}-1} dt_1 dt_2, \tag{5.25}
$$

and the domain $\mathcal{D} := \{(t_1, t_2) \mid t_1, t_2 > 0, t_1+t_2 < 1\}$ *is a simplex.*

Remark 5.2. *Other forms for the cross moment of the feasible GRR estimators are given in Srivastava and Chaturvedi (1983) and Jimichi (1999). These are obtained by applying the negative binomial expansion*

$$
\frac{1}{(1-t)^r} = \sum_{g=0}^{\infty} \binom{g+r-1}{g} t^g
$$

to $\mathcal{B}_{l_1+1\,l_2+1}(\cdot, \cdot, \cdot)$ *in (5.25). However, it is difficult to check the convergence and numerically evaluate these forms.*

For the first and second moments, we have the following questions:

(Q3) Are summation and expectation in (5.23) exchangeable?
(Q4) Does the series in (5.24) converge?

We begin addressing question (Q3) by preparing some lemmas.

Lemma 5.4. $\widehat{\alpha}_{iL}^*$ converges in quadratic mean to the feasible GRR estimator $\widehat{\alpha}_i^*$:
$$\lim_{L \to \infty} E\left(|\widehat{\alpha}_{iL}^* - \widehat{\alpha}_i^*|^2\right) = 0.$$

Proof. By using (5.5) and (5.16), we have
$$\widehat{\alpha}_{iL}^* - \widehat{\alpha}_i^* = -\widehat{\alpha}_i^* \left(\frac{\nu-1}{\nu}\right)^{L+1} \left(\frac{V}{Z_i^2 + V}\right)^{L+1}, \qquad (5.26)$$

and
$$|\widehat{\alpha}_{iL}^* - \widehat{\alpha}_i^*|^2 = |\widehat{\alpha}_i^*|^2 \left(\frac{\nu-1}{\nu}\right)^{2(L+1)} \left(\frac{V}{Z_i^2 + V}\right)^{2(L+1)} \leq |\widehat{\alpha}_i^*|^2 \left(\frac{\nu-1}{\nu}\right)^{2(L+1)}.$$

Hence, by using Proposition 5.1,
$$\lim_{L \to \infty} E\left(|\widehat{\alpha}_{iL}^* - \widehat{\alpha}_i^*|^2\right) \leq E\left(|\widehat{\alpha}_i^*|^2\right) \lim_{L \to \infty} \left\{\left(\frac{\nu-1}{\nu}\right)^2\right\}^{L+1} = 0.$$
□

Lemma 5.5. $\widehat{\alpha}_{iL_1}^* \widehat{\alpha}_{jL_2}^*$ converges in first mean to $\widehat{\alpha}_i^* \widehat{\alpha}_j^*$:
$$\lim_{L_1, L_2 \to \infty} E\left(|\widehat{\alpha}_{iL_1}^* \widehat{\alpha}_{jL_2}^* - \widehat{\alpha}_i^* \widehat{\alpha}_j^*|\right) = 0.$$

Proof. By using
$$|\widehat{\alpha}_{iL_1}^* \widehat{\alpha}_{jL_2}^* - \widehat{\alpha}_i^* \widehat{\alpha}_j^*| \leq |\widehat{\alpha}_{iL_1}^* - \widehat{\alpha}_i^*||\widehat{\alpha}_{jL_2}^* - \widehat{\alpha}_j^*| + |\widehat{\alpha}_j^*||\widehat{\alpha}_{iL_1}^* - \widehat{\alpha}_i^*| + |\widehat{\alpha}_{iL_1}^*||\widehat{\alpha}_{jL_2}^* - \widehat{\alpha}_j^*|$$
and the Cauchy-Schwarz inequality, we have
$$E\left(|\widehat{\alpha}_{iL_1}^* \widehat{\alpha}_{jL_2}^* - \widehat{\alpha}_i^* \widehat{\alpha}_j^*|\right) \leq \left\{E\left(|\widehat{\alpha}_{iL_1}^* - \widehat{\alpha}_i^*|^2\right)\right\}^{\frac{1}{2}} \left\{E\left(|\widehat{\alpha}_{jL_2}^* - \widehat{\alpha}_j^*|^2\right)\right\}^{\frac{1}{2}}$$
$$+ \left\{E\left(|\widehat{\alpha}_j^*|^2\right)\right\}^{\frac{1}{2}} \left\{E\left(|\widehat{\alpha}_{iL_1}^* - \widehat{\alpha}_i^*|^2\right)\right\}^{\frac{1}{2}}$$
$$+ \left\{E\left(|\widehat{\alpha}_i^*|^2\right)\right\}^{\frac{1}{2}} \left\{E\left(|\widehat{\alpha}_{jL_2}^* - \widehat{\alpha}_j^*|^2\right)\right\}^{\frac{1}{2}}.$$

Hence, from Lemma 5.4, the right-hand side in the above inequality converges to 0 as L_1 and L_2 go to infinity. The proof is complete. □

Theorem 5.3 gives an answer to question (Q3) as follows:

Theorem 5.3. Let $\nu \geq 2$. Then,
$$E(\widehat{\alpha}_i^* \widehat{\alpha}_j^*) = \lim_{L_1, L_2 \to \infty} E(\widehat{\alpha}_{iL_1}^* \widehat{\alpha}_{jL_2}^*)$$
$$= \sigma_{\widehat{\alpha}_i} \sigma_{\widehat{\alpha}_j} \lim_{L_1, L_2 \to \infty} \sum_{\ell_1=0}^{L_1} \sum_{\ell_2=0}^{L_2} \left(\frac{\nu-1}{\nu}\right)^{\ell_1+\ell_2} E\left(\frac{Z_i^3 V^{\ell_1}}{(Z_i^2+V)^{\ell_1+1}} \frac{Z_j^3 V^{\ell_2}}{(Z_j^2+V)^{\ell_2+1}}\right).$$
(5.27)

Proof. The proof is obvious by Lemma 5.5 and

$$\left|\mathrm{E}(\widehat{\alpha}^*_{iL_1}\widehat{\alpha}^*_{jL_2}) - \mathrm{E}(\widehat{\alpha}^*_i \widehat{\alpha}^*_j)\right| = \left|\mathrm{E}(\widehat{\alpha}^*_{iL_1}\widehat{\alpha}^*_{jL_2} - \widehat{\alpha}^*_i \widehat{\alpha}^*_j)\right| \leq \mathrm{E}\left(|\widehat{\alpha}^*_{iL_1}\widehat{\alpha}^*_{jL_2} - \widehat{\alpha}^*_i \widehat{\alpha}^*_j|\right).$$

□

Next, we prepare a lemma for question (Q4).

Lemma 5.6. *Let $\nu \geq 5$. Then, $\mathcal{B}_{\ell_1+1\ \ell_2+1}(\cdot,\cdot,\cdot)$ in (5.24) is evaluated for all $m_1, m_2, \ell_1, \ell_2 \geq 0$ as follows:*

$$\mathcal{B}_{\ell_1+1\ \ell_2+1}\left(m_1 + \frac{5}{2}, m_2 + \frac{5}{2}, \ell_1 + \ell_2 + \frac{\nu}{2}\right) < B\left(m_1 + \frac{5}{2}, m_2 + \frac{5}{2}, \frac{\nu}{2} - 2\right).$$

Proof. Note that for $(t_1, t_2) \in \mathcal{D}$,

$$\frac{1 - t_1 - t_2}{1 - t_1} = 1 - \frac{t_2}{1 - t_1} < 1, \quad \frac{1 - t_1 - t_2}{1 - t_2} = 1 - \frac{t_1}{1 - t_2} < 1.$$

From definition (5.25) of $\mathcal{B}_{\ell_1+1\ \ell_2+1}(\cdot,\cdot,\cdot)$, for all $m_1, m_2, \ell_1, \ell_2 \geq 0$,

$$\mathcal{B}_{\ell_1+1\ \ell_2+1}\left(m_1 + \frac{5}{2}, m_2 + \frac{5}{2}, \ell_1 + \ell_2 + \frac{\nu}{2}\right)$$

$$< \iint_{\mathcal{D}} t_1^{m_1+\frac{5}{2}-1} t_2^{m_2+\frac{5}{2}-1}(1 - t_1 - t_2)^{(\frac{\nu}{2}-2)-1} dt_1 dt_2$$

$$= B\left(m_1 + \frac{5}{2}, m_2 + \frac{5}{2}, \frac{\nu}{2} - 2\right). \tag{5.28}$$

The proof is complete. □

The convergence of the cross moment is guaranteed by the following theorem:

Theorem 5.4. *Let $\delta_i, \delta_j \geq 0$ and $\nu \geq 5$. Then, the following series converges:*

$$\mu_{11}(\delta_i, \delta_j, \nu) := \sum_{\ell_1=0}^{\infty} \sum_{\ell_2=0}^{\infty} \left(\frac{\nu-1}{\nu}\right)^{\ell_1+\ell_2}$$

$$\times \sum_{m_1=0}^{\infty} \sum_{m_2=0}^{\infty} \frac{\mathcal{B}_{\ell_1+1\ \ell_2+1}\left(m_1 + \frac{5}{2}, m_2 + \frac{5}{2}, \ell_1 + \ell_2 + \frac{\nu}{2}\right)}{B\left(m_1 + \frac{3}{2}, m_2 + \frac{3}{2}, \frac{\nu}{2}\right)} p(m_1; \delta_i) p(m_2; \delta_j).$$

(5.29)

Proof. By using basic properties of the gamma function, we have

$$B\left(m_1 + \frac{5}{2}, m_2 + \frac{5}{2}, \frac{\nu}{2} - 2\right) = \frac{\Gamma\left(m_1 + \frac{5}{2}\right)\Gamma\left(m_2 + \frac{5}{2}\right)\Gamma\left(\frac{\nu}{2} - 2\right)}{\Gamma\left(m_1 + m_2 + \frac{\nu}{2} + 3\right)}$$

$$= \left(m_1 + \frac{3}{2}\right)\left(m_2 + \frac{3}{2}\right) \frac{\Gamma\left(m_1 + \frac{3}{2}\right)\Gamma\left(m_2 + \frac{3}{2}\right)\Gamma\left(\frac{\nu}{2} - 2\right)}{\Gamma\left(m_1 + m_2 + \frac{\nu}{2} + 3\right)},$$

$$\frac{\Gamma\left(\frac{\nu}{2} - 2\right)}{\Gamma\left(\frac{\nu}{2}\right)} \leq \frac{4}{3}, \ \nu \geq 5.$$

The series is bounded for any $\nu \geq 5$ and $\delta_i, \delta_j > 0$ by Lemma 5.6:

$$\mu_{11}(\delta_i, \delta_j, \nu) < \sum_{\ell_1=0}^{\infty} \sum_{\ell_2=0}^{\infty} \left(\frac{\nu-1}{\nu}\right)^{\ell_1+\ell_2}$$

$$\times \sum_{m_1=0}^{\infty} \sum_{m_2=0}^{\infty} \frac{B\left(m_1 + \frac{5}{2}, m_2 + \frac{5}{2}, \frac{\nu}{2} - 2\right)}{B\left(m_1 + \frac{3}{2}, m_2 + \frac{3}{2}, \frac{\nu}{2}\right)} p(m_1; \delta_i) p(m_2; \delta_j)$$

$$= \sum_{\ell_1=0}^{\infty} \sum_{\ell_2=0}^{\infty} \left(\frac{\nu-1}{\nu}\right)^{\ell_1+\ell_2}$$

$$\times \sum_{m_1=0}^{\infty} \sum_{m_2=0}^{\infty} \left(m_1 + \frac{3}{2}\right) \left(m_2 + \frac{3}{2}\right) \frac{\Gamma\left(\frac{\nu}{2} - 2\right)}{\Gamma\left(\frac{\nu}{2}\right)} p(m_1; \delta_i) p(m_2; \delta_j)$$

$$\leq \frac{4}{3}\nu^2 \left(\delta_i + \frac{3}{2}\right)\left(\delta_j + \frac{3}{2}\right) = \frac{\nu^2}{3}(2\delta_i + 3)(2\delta_j + 3) < \infty. \tag{5.30}$$

Hence, the series $\mu_{11}(\delta_i, \delta_j, \nu)$ converges because it is positive. $\square$

Question (Q4) has been answered by Theorem 5.4.

Remark 5.3. *Note that if* $\alpha_i = 0$ *or* $\alpha_j = 0$, *then*

$$E\left(\widehat{\alpha}_i^* \widehat{\alpha}_j^*\right) = 0.$$

5.4 Mean Squared Error Criteria

The vector of the first moments for the feasible GRR estimators is obtained by using (5.11) as follows:

$$E(\widehat{\boldsymbol{\alpha}}_p^*) = \begin{bmatrix} E(\widehat{\alpha}_1^*) \\ \vdots \\ E(\widehat{\alpha}_p^*) \end{bmatrix} = \begin{bmatrix} \mu_{11}\alpha_1 \\ \vdots \\ \mu_{1p}\alpha_p \end{bmatrix} = \begin{bmatrix} \alpha_1 & & \mathbf{O} \\ & \ddots & \\ \mathbf{O} & & \alpha_p \end{bmatrix} \begin{bmatrix} \mu_{11} \\ \vdots \\ \mu_{1p} \end{bmatrix} =: \mathcal{A}\boldsymbol{\mu}_1, \tag{5.31}$$

where

$$\mu_{1i} := \mu_1(\delta_i, \nu) = \sum_{\ell=0}^{\infty}\left(\frac{\nu-1}{\nu}\right)^{\ell} \sum_{m=0}^{\infty} \frac{B\left(\ell + \frac{\nu}{2}, m + \frac{5}{2}\right)}{B\left(\frac{\nu}{2}, m + \frac{3}{2}\right)} p(m; \delta_i).$$

(See also (5.20).) Thus, the bias vector of $\widehat{\boldsymbol{\alpha}}_p^*$ is obtained by

$$\text{bias}(\widehat{\boldsymbol{\alpha}}_p^*) := E(\widehat{\boldsymbol{\alpha}}_p^*) - \boldsymbol{\alpha}_p = \mathcal{A}\widetilde{\boldsymbol{\mu}}_1, \tag{5.32}$$

where $\widetilde{\boldsymbol{\mu}}_1 := [\mu_{11} - 1, \ldots, \mu_{1p} - 1]'$.

5.4. MEAN SQUARED ERROR CRITERIA   61

Next, we consider the variance-covariance matrix of the feasible GRR estimators. We know the variance of $\widehat{\alpha}_i^*$ by using (5.11) and (5.12) as follows:

$$V(\widehat{\alpha}_i^*) = E(\widehat{\alpha}_i^{*2}) - E(\widehat{\alpha}_i^*)^2 = \alpha_i^2 \mu_{2i} - \alpha_i^2 \mu_{1i}^2, \tag{5.33}$$

where

$$\mu_{2i} := \mu_2(\delta_i, \nu) = \frac{1}{2\delta_i} \sum_{\ell=0}^{\infty} (\ell+1) \left(\frac{\nu-1}{\nu}\right)^{\ell} \sum_{m=0}^{\infty} (2m+1) \frac{B\left(\ell + \frac{\nu}{2}, m + \frac{7}{2}\right)}{B\left(\frac{\nu}{2}, m + \frac{3}{2}\right)} p(m; \delta_i).$$

(See also (5.21).) We have the covariance between $\widehat{\alpha}_i^*$ and $\widehat{\alpha}_j^*$ by using Theorem 5.2 as follows:

$$\text{Cov}(\widehat{\alpha}_i^*, \widehat{\alpha}_j^*) = E(\widehat{\alpha}_i^* \widehat{\alpha}_j^*) - E(\widehat{\alpha}_i^*)E(\widehat{\alpha}_j^*) = \alpha_i \alpha_j \mu_{11ij} - \alpha_i \alpha_j \mu_{1i} \mu_{1j}, \tag{5.34}$$

where

$$\mu_{11ij} := \mu_{11}(\delta_i, \delta_j, \nu)$$
$$= \sum_{\ell_1=0}^{\infty} \sum_{\ell_2=0}^{\infty} \left(\frac{\nu-1}{\nu}\right)^{\ell_1+\ell_2}$$
$$\times \sum_{m_1=0}^{\infty} \sum_{m_2=0}^{\infty} \frac{\mathcal{B}_{\ell_1+1\,\ell_2+1}\left(m_1 + \frac{5}{2}, m_2 + \frac{5}{2}, \ell_1 + \ell_2 + \frac{\nu}{2}\right)}{B\left(m_1 + \frac{3}{2}, m_2 + \frac{3}{2}, \frac{\nu}{2}\right)} p(m_1; \delta_i) p(m_2; \delta_j).$$

(See also (5.29).) Therefore, the variance-covariance matrix is obtained by (5.33) and (5.34) as follows:

$$V(\widehat{\boldsymbol{\alpha}}_p^*) = \begin{bmatrix} V(\widehat{\alpha}_1^*) & \cdots & \text{Cov}(\widehat{\alpha}_1^*, \widehat{\alpha}_p^*) \\ \vdots & \ddots & \vdots \\ \text{Cov}(\widehat{\alpha}_p^*, \widehat{\alpha}_1^*) & \cdots & V(\widehat{\alpha}_p^*) \end{bmatrix}$$
$$= \begin{bmatrix} \alpha_1^2 \mu_{21} - \alpha_1^2 \mu_{11}^2 & \cdots & \alpha_1 \alpha_p \mu_{111p} - \alpha_1 \alpha_p \mu_{11} \mu_{1p} \\ \vdots & \ddots & \vdots \\ \alpha_p \alpha_1 \mu_{11p1} - \alpha_p \alpha_1 \mu_{1p} \mu_{11} & \cdots & \alpha_p^2 \mu_{2p} - \alpha_p^2 \mu_{1p}^2 \end{bmatrix}$$
$$= \mathcal{A}(\mathbf{M}_2 - \boldsymbol{\mu}_1 \boldsymbol{\mu}_1')\mathcal{A}, \tag{5.35}$$

where

$$\mathbf{M}_2 := \begin{bmatrix} \mu_{21} & \cdots & \mu_{111p} \\ \vdots & \ddots & \vdots \\ \mu_{11p1} & \cdots & \mu_{2p} \end{bmatrix}.$$

Hence, the MSE matrix of the feasible GRR estimator $\widehat{\boldsymbol{\alpha}}_p^*$ is obtained by (5.32) and (5.35) as follows:

$$\text{MSE}(\widehat{\boldsymbol{\alpha}}_p^*) = V(\widehat{\boldsymbol{\alpha}}_p^*) + \text{bias}(\widehat{\boldsymbol{\alpha}}_p^*)\text{bias}(\widehat{\boldsymbol{\alpha}}_p^*)'$$
$$= \mathcal{A}(\mathbf{M}_2 - \boldsymbol{\mu}_1 \boldsymbol{\mu}_1')\mathcal{A} + \mathcal{A}\widetilde{\boldsymbol{\mu}}_1 \widetilde{\boldsymbol{\mu}}_1' \mathcal{A}$$
$$= \mathcal{A}(\mathbf{M}_2 - \boldsymbol{\mu}_1 \boldsymbol{\mu}_1' + \widetilde{\boldsymbol{\mu}}_1 \widetilde{\boldsymbol{\mu}}_1')\mathcal{A}. \tag{5.36}$$

Recall that the MSE of $\widehat{\alpha}_i^*$ and the MCE of $\widehat{\alpha}_i^*$ and $\widehat{\alpha}_j^*$ are defined by

$$\operatorname{MSE}\left(\widehat{\alpha}_i^*\right) := \operatorname{E}\left(\widehat{\alpha}_i^* - \alpha_i\right)^2, \tag{5.37}$$

$$\operatorname{MCE}\left(\widehat{\alpha}_i^*, \widehat{\alpha}_j^*\right) := \operatorname{E}\left(\widehat{\alpha}_i^* - \alpha_i\right)\left(\widehat{\alpha}_j^* - \alpha_j\right), \tag{5.38}$$

respectively. These are the (i,j) elements of the MSE matrix of $\widehat{\boldsymbol{\alpha}}^*$ in (5.35):

$$\left[\operatorname{MSE}\left(\widehat{\boldsymbol{\alpha}}_p^*\right)\right]_{ij} = \begin{cases} \alpha_i^2\left(\mu_{2i} - 2\mu_{1i} + 1\right) = \operatorname{MSE}\left(\widehat{\alpha}_i^*\right) & ; \text{if } i = j, \\ \alpha_i\alpha_j\left\{\mu_{11ij} - (\mu_{1i} + \mu_{1j}) + 1\right\} = \operatorname{MCE}\left(\widehat{\alpha}_i^*, \widehat{\alpha}_j^*\right) & ; \text{if } i \neq j. \end{cases} \tag{5.39}$$

The relative mean squared error (RMSE) of $\widehat{\alpha}_i^*$ and the relative mean cross error (RMCE) of $\widehat{\alpha}_i^*$ and $\widehat{\alpha}_j^*$ are defined by

$$\operatorname{RMSE}\left(\widehat{\alpha}_i^*\right) := \frac{\operatorname{MSE}\left(\widehat{\alpha}_i^*\right)}{\alpha_i^2}, \tag{5.40}$$

$$\operatorname{RMCE}\left(\widehat{\alpha}_i^*, \widehat{\alpha}_j^*\right) := \frac{\operatorname{MCE}\left(\widehat{\alpha}_i^*, \widehat{\alpha}_j^*\right)}{\alpha_i\alpha_j}. \tag{5.41}$$

The RMSE matrix of $\widehat{\boldsymbol{\alpha}}^*$ is defined by

$$\operatorname{RMSE}\left(\widehat{\boldsymbol{\alpha}}_p^*\right) := \mathcal{A}^{-1}\operatorname{MSE}\left(\widehat{\boldsymbol{\alpha}}_p^*\right)\mathcal{A}^{-1} = \mathbf{M}_2 - \boldsymbol{\mu}_1\boldsymbol{\mu}_1' + \widetilde{\boldsymbol{\mu}}_1\widetilde{\boldsymbol{\mu}}_1', \tag{5.42}$$

and the (i,j) element of the RMSE matrix of $\widehat{\boldsymbol{\alpha}}^*$ is called the RMSE or the RMCE as follows:

$$\left[\operatorname{RMSE}\left(\widehat{\boldsymbol{\alpha}}_p^*\right)\right]_{ij} = \begin{cases} \mu_{2i} - 2\mu_{1i} + 1 = \operatorname{RMSE}\left(\widehat{\alpha}_i^*\right) & ; \text{if } i = j, \\ \mu_{11ij} - (\mu_{1i} + \mu_{1j}) + 1 = \operatorname{RMCE}\left(\widehat{\alpha}_i^*, \widehat{\alpha}_j^*\right) & ; \text{if } i \neq j. \end{cases} \tag{5.43}$$

Furthermore, the efficiency of the OLS estimator $\widehat{\alpha}_i$ relative to the feasible GRR estimator $\widehat{\alpha}_i^*$ is defined by

$$\operatorname{REff}(\widehat{\alpha}_i^*, \widehat{\alpha}_i) := \frac{\operatorname{MSE}(\widehat{\alpha}_i^*)}{\operatorname{MSE}(\widehat{\alpha}_i)} = \frac{\operatorname{MSE}(\widehat{\alpha}_i^*)}{\operatorname{V}(\widehat{\alpha}_i)} = \tau_i^2\operatorname{RMSE}(\widehat{\alpha}_i^*) = 2\delta_i\operatorname{RMSE}(\widehat{\alpha}_i^*). \tag{5.44}$$

Note that this is called the *relative efficiency*, and

$$\operatorname{REff}(\widehat{\alpha}_i^*, \widehat{\alpha}_i) < 1 \iff \operatorname{MSE}(\widehat{\alpha}_i^*) < \operatorname{MSE}(\widehat{\alpha}_i). \tag{5.45}$$

Finally, the TMSE of $\widehat{\boldsymbol{\alpha}}^*$ is given by the trace of the MSE matrix $\operatorname{MSE}\left(\widehat{\boldsymbol{\alpha}}^*\right)$:

$$\operatorname{TMSE}\left(\widehat{\boldsymbol{\alpha}}_p^*\right) = \operatorname{trace} \operatorname{MSE}\left(\widehat{\boldsymbol{\alpha}}_p^*\right) = \sum_{i=1}^p \alpha_i^2\left(\mu_{2i} - 2\mu_{1i} + 1\right). \tag{5.46}$$

Remark 5.4. *From Remark 5.1, if $\alpha_i = 0$, then the variance of the feasible GRR estimator $\widehat{\alpha}_i^*$ is equal to its MSE:*

$$V(\widehat{\alpha}_i^*) = MSE(\widehat{\alpha}_i^*) = \frac{\sigma^2}{\lambda_i}\widetilde{\mu}_2(\nu),$$

where $\widetilde{\mu}_2(\nu)$ is defined by (5.22). Therefore, the efficiency of the OLS estimator $\widehat{\alpha}_i$ relative to the feasible GRR estimator $\widehat{\alpha}_i^*$ is given by

$$REff(\widehat{\alpha}_i^*, \widehat{\alpha}_i) = \widetilde{\mu}_2(\nu). \tag{5.47}$$

5.5 Numerical Evaluations

5.5.1 First and Second Moments

It is not difficult to directly compute the series $\mu_1(\delta_i, \nu)$, $\mu_2(\delta_i, \nu)$ in the first and second moments in recent computer environments. To numerically evaluate them, we use the following finite series:

$$\mu_1(L, M, \delta_i, \nu) := \sum_{\ell=0}^{L} \left(\frac{\nu-1}{\nu}\right)^\ell \sum_{m=0}^{M} \frac{B\left(\ell + \frac{\nu}{2}, m + \frac{5}{2}\right)}{B\left(\frac{\nu}{2}, m + \frac{3}{2}\right)} p(m; \delta_i), \tag{5.48}$$

$$\mu_2(L, M, \delta_i, \nu) := \frac{1}{2\delta_i} \sum_{\ell=0}^{L} (\ell+1) \left(\frac{\nu-1}{\nu}\right)^\ell \sum_{m=0}^{M} (2m+1) \frac{B\left(\ell + \frac{\nu}{2}, m + \frac{7}{2}\right)}{B\left(\frac{\nu}{2}, m + \frac{3}{2}\right)} p(m; \delta_i). \tag{5.49}$$

When computing the finite series, we can accelerate the calculation by making use of some recurrence relations with respect to the beta function ratio. For example, we use the following recurrence relations for $\mu_1(L, M, \delta_i, \nu)$:

$$\frac{B\left((\ell+1) + \frac{\nu}{2}, m + \frac{5}{2}\right)}{B\left(\frac{\nu}{2}, m + \frac{3}{2}\right)} = \frac{\ell + \frac{\nu}{2}}{\ell + m + \frac{\nu+5}{2}} \frac{B\left(\ell + \frac{\nu}{2}, m + \frac{5}{2}\right)}{B\left(\frac{\nu}{2}, m + \frac{3}{2}\right)},$$

$$\frac{B\left(\ell + \frac{\nu}{2}, (m+1) + \frac{5}{2}\right)}{B\left(\frac{\nu}{2}, (m+1) + \frac{3}{2}\right)} = \frac{\left(m + \frac{5}{2}\right)\left(m + \frac{\nu+3}{2}\right)}{\left(\frac{\nu+3}{2}\right)\left(\ell + m + \frac{\nu+5}{2}\right)} \frac{B\left(\ell + \frac{\nu}{2}, m + \frac{5}{2}\right)}{B\left(\frac{\nu}{2}, m + \frac{3}{2}\right)}.$$

Figure 5.2 shows the convergence of $\mu_1(L, M, \delta_i, \nu)$ and $\mu_2(L, M, \delta_i, \nu)$ as $L = M = 1, \ldots, 100$ for the case $(\delta_i, \nu) = (50, 100)$. The limit value of $\mu_1(L, M, 50, 100)$ is 0.9900042, attained for $L = M \geq 90$, and that of $\mu_2(L, M, 50, 100)$ is 0.9903110, attained for $L = M \geq 92$.

5.5.2 Cross Moment

The calculation of cross moment (5.24) is essentially equivalent to the calculation of the series $\mu_{11}(\delta_i, \delta_j, \nu)$ in (5.29). To numerically evaluate the

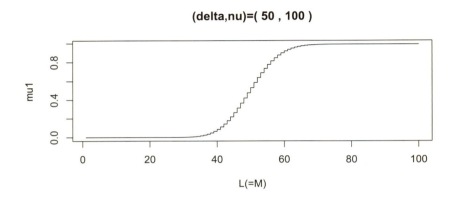

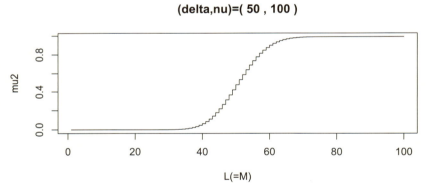

Figure 5.2: $\mu_1(L, M, \delta_i, \nu)$ and $\mu_2(L, M, \delta_i, \nu)$ versus $L(= M) = 1, \ldots, 100$ if $(\delta_i, \nu) = (50, 100)$

cross moment, we need to use the following finite series:

$$\mu_{11}(L_1, L_2, M_1, M_2, \delta_i, \delta_j, \nu)$$
$$:= \sum_{\ell_1=0}^{L_1} \sum_{\ell_2=0}^{L_2} \left(\frac{\nu-1}{\nu}\right)^{\ell_1+\ell_2}$$
$$\times \sum_{m_1=0}^{M_1} \sum_{m_2=0}^{M_2} \frac{\mathcal{B}_{\ell_1+1\ \ell_2+1}\left(m_1+\frac{5}{2}, m_2+\frac{5}{2}, \ell_1+\ell_2+\frac{\nu}{2}\right)}{B\left(m_1+\frac{3}{2}, m_2+\frac{3}{2}, \frac{\nu}{2}\right)} p(m_1; \delta_i) p(m_2; \delta_j).$$
(5.50)

First of all, from the properties of the beta function, we know

$$B\left(m_1+\frac{3}{2}, m_2+\frac{3}{2}, \frac{\nu}{2}\right) = B\left(m_1+\frac{3}{2}, m_2+\frac{3}{2}\right) B\left(m_1+m_2+3, \frac{\nu}{2}\right).$$

Therefore, we need only calculate the beta functions. Next, we consider numerical evaluation of $\mathcal{B}_{\ell_1+1\ \ell_2+1}(\cdot, \cdot, \cdot)$. (See also (5.25).) There are many methods for numerical integration, but let us use the *method of good lattice points* (GLP) (e.g. Sugihara and Murota (1994)) to numerically integrate the integrand

$$f(t_1, t_2) := \frac{1}{(1-t_1)^{\ell_2+1}} \frac{1}{(1-t_2)^{\ell_1+1}} t_1^{m_1+\frac{5}{2}-1} t_2^{m_2+\frac{5}{2}-1} (1-t_1-t_2)^{\ell_1+\ell_2+\frac{\nu}{2}-1}$$

in $\mathcal{B}_{\ell_1+1\ \ell_2+1}(\cdot, \cdot, \cdot)$ on $\mathcal{D}$. Figure 5.3 is a plot of $f(t_1, t_2)$ when $m_1 = m_2 = \ell_1 = \ell_2 = 0$, $\nu = 5$.

We use the following approximation formula for $\mathcal{B}_{\ell_1+1\ \ell_2+1}$:

$$B_N := \frac{1}{N} \sum_{k \in K} f\left(\phi_q\left(\left\{\frac{g_1^{(N)}}{N}k\right\}\right), \phi_q\left(\left\{\frac{g_2^{(N)}}{N}k\right\}\right)\right) \phi_q'\left(\left\{\frac{g_1^{(N)}}{N}k\right\}\right) \phi_q'\left(\left\{\frac{g_2^{(N)}}{N}k\right\}\right),$$

where $K(\subset \{0, 1, \ldots, N-1\})$ is a set such that

$$\left(\phi_q\left(\left\{\frac{g_1^{(N)}}{N}k\right\}\right), \phi_q\left(\left\{\frac{g_2^{(N)}}{N}k\right\}\right)\right) \in \mathcal{D} = \{(t_1, t_2) \mid t_1, t_2 > 0, t_1 + t_2 < 1\},$$

and $\{x\}$ is the decimal part of x [3]. Additionally,

$$\phi_q(x) := \frac{(2q+1)!}{q! q!} \int_0^x u^q (1-u)^q du,$$

and

$$N = F_n, \quad g_1^{(N)} = 1, \quad g_2^{(N)} = F_{n-1},$$

---
[3] Note that k is not the ridge coefficient.

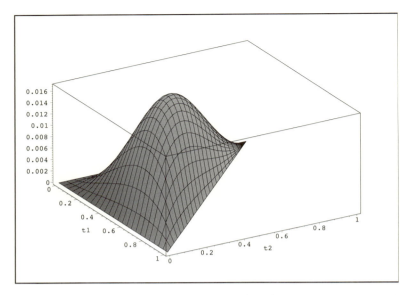

Figure 5.3: Integrand $f(t_1, t_2)$ in $\mathcal{B}_{\ell_1+1\ \ell_2+1}(\cdot, \cdot, \cdot)$: $m_1 = m_2 = \ell_1 = \ell_2 = 0$, $\nu = 5$.

and F_n is the Fibonacci sequence defined by

$$F_1 = F_2 = 1, \quad F_{n+2} = F_{n+1} + F_n, \quad n = 1, 2, \ldots$$

Table 5.1 gives the values of $(N, g_1^{(N)}, g_2^{(N)})$. When values $N = 4181$ ($n = 19$) and $q = 2$ are adopted, the lattice points $\left(\phi_q\left(\left\{g_1^{(N)}k/N\right\}\right), \phi_q\left(\left\{g_2^{(N)}k/N\right\}\right)\right)$ on the simplex $\mathcal{D}$ are as represented in Figure 5.4.

Let $q = 2$, $N = 4181$ and $m_1 = m_2 = \ell_1 = \ell_2 = 0$, $\nu = 5$. We obtain the following result by the GLP method:

$$\mathcal{B}_{1\ 1}\left(\frac{5}{2}, \frac{5}{2}, \frac{5}{2}\right) = \iint_{\mathcal{D}} \frac{1}{(1-t_1)(1-t_2)} t_1^{\frac{3}{2}} t_2^{\frac{3}{2}} (1 - t_1 - t_2)^{\frac{3}{2}} dt_1 dt_2 = 0.00316429.$$

Let us choose the special case of $(\delta_i, \delta_j, \nu) = (1, 1, 5)$, and consider the convergence of $\mu_{11}(L_1, L_2, M_1, M_2, \delta_i, \delta_j, \nu)$. Table 5.2 lists the values of $\mu_{11}(L_1, L_2, M_1, M_2, \delta_i, \delta_j, \nu)$ for the cases $n = 18, 19, \ldots, 22$, $L_1 = L_2 = 20, 21, \ldots, 25, 30, 40, 50$, $M_1 = M_2 = 9, 10, 11$. The table implies the following result [4]:

$$\mu_{11}(L_1, L_2, M_1, M_2, 1, 1, 5) = 0.6100.$$

---

[4] We used the parallel programming environments "LAM/MPI" and "Rmpi" on the grid system "Globus Toolkit 3.2" to compute the cross moment.

5.5. NUMERICAL EVALUATIONS

Table 5.1: Table of the set $(n, N, g_1^{(N)}, g_2^{(N)})$

n	$N(=F_n)$	$g_1^{(N)}$	$g_2^{(N)}$
10	55	1	34
11	89	1	55
12	144	1	89
13	233	1	144
14	377	1	233
15	610	1	377
16	987	1	610
17	1597	1	987
18	2584	1	1597
19	4181	1	2584

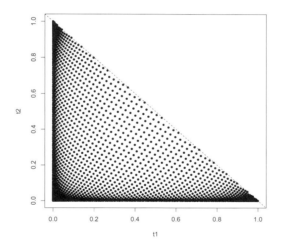

Figure 5.4: 2091 lattice points $\left(\phi_p\left(\left\{\frac{g_1^{(N)}}{N}k\right\}\right), \phi_p\left(\left\{\frac{g_2^{(N)}}{N}k\right\}\right) \right)$ on simplex $\mathcal{D}$

68   5 EXACT MOMENTS OF FEASIBLE GRR ESTIMATOR

Table 5.2: Numerical results for $\mu_{11}(L_1, L_2, M_1, M_2, 1, 1, 5)$ for the cases $n = 18, 19, \ldots, 22$, $L = L_1 = L_2 = 20, 21, \ldots, 25, 30, 40, 50$, and $M = M_1 = M_2 = 9, 10, 11$.

$n = 18 (N = 2584)$

$M \backslash L$	20	21	22	23	24	25	30	40	50
9	0.6099484	0.6099612	0.6099704	0.6099770	0.6099818	0.6099852	0.6099926	0.6099946	0.6099947
10	0.6099485	0.6099614	0.6099706	0.6099772	0.6099819	0.6099854	0.6099927	0.6099947	0.6099948
11	0.6099486	0.6099614	0.6099706	0.6099772	0.6099820	0.6099854	0.6099928	0.6099947	0.6099948

$n = 19$ $(N = 4181)$

$M \backslash L$	20	21	22	23	24	25	30	40	50
9	0.6099862	0.6099991	0.6100082	0.6100148	0.6100196	0.6100230	0.6100304	0.6100324	0.6100325
10	0.6099863	0.6099992	0.6100084	0.6100150	0.6100198	0.6100232	0.6100306	0.6100325	0.6100326
11	0.6099864	0.6099992	0.6100084	0.6100150	0.6100198	0.6100232	0.6100306	0.6100325	0.6100326

$n = 20$ $(N = 6765)$

$M \backslash L$	20	21	22	23	24	25	30	40	50
9	0.6098999	0.6099128	0.6099220	0.6099286	0.6099333	0.6099368	0.6099441	0.6099461	0.6099462
10	0.6099001	0.6099129	0.6099221	0.6099287	0.6099335	0.6099369	0.6099443	0.6099462	0.6099463
11	0.6099001	0.6099129	0.6099221	0.6099287	0.6099335	0.6099369	0.6099443	0.6099462	0.6099464

$n = 21$ $(N = 10946)$

$M \backslash L$	20	21	22	23	24	25	30	40	50
9	0.6099860	0.6099989	0.6100081	0.6100147	0.6100194	0.6100229	0.6100302	0.6100322	0.6100323
10	0.6099862	0.6099990	0.6100082	0.6100148	0.6100196	0.6100230	0.6100304	0.6100323	0.6100325
11	0.6099862	0.6099991	0.6100082	0.6100148	0.6100196	0.6100230	0.6100304	0.6100324	0.6100325

$n = 22$ $(N = 17711)$

$M \backslash L$	20	21	22	23	24	25	30	40	50
9	0.6099826	0.6099955	0.6100047	0.6100113	0.6100160	0.6100195	0.6100269	0.6100288	0.6100289
10	0.6099828	0.6099957	0.6100048	0.6100114	0.6100162	0.6100196	0.6100270	0.6100290	0.6100291
11	0.6099828	0.6099957	0.6100049	0.6100115	0.6100162	0.6100197	0.6100270	0.6100290	0.6100291

Table 5.3: Exact moments and their Monte Carlo approximations

	$\mu_1(\delta_i,\nu)$	$\mu_2(\delta_i,\nu)$	$\mu_{11}(\delta_i,\delta_j,\nu)$	$E(\widehat{\alpha}_i^*)$	$E(\widehat{\alpha}_i^{*2})$	$E(\widehat{\alpha}_i^*\widehat{\alpha}_j^*)$
Exact	0.7716528	1.038585	0.6100	1.543306	4.154339	2.4400
Monte Carlo	0.7718257	1.038736	0.6041	1.543651	4.154944	2.4166

Remark 5.5. *If we take large values of (δ_i,δ_j,ν) (for example, $(\delta_i,\delta_j,\nu) = (50,50,100)$), then a more powerful computing environment is needed to compute the cross moment because of the properties of the series $\left(\frac{\nu-1}{\nu}\right)^{\ell_i}$ and $p(m_i;\delta_i)$ $(i=1,2)$. Thus, we select the special case of $(\delta_i,\delta_j,\nu) = (1,1,5)$.*

5.5.3 Monte Carlo Simulations

Let us compare the exact moments of the feasible GRR estimator with some Monte Carlo results. (See Table 5.3.)

We set
$$\alpha_i = \alpha_j = 2, \quad \sigma = 1, \quad \lambda_i = \lambda_j = \frac{1}{2}, \quad \nu = 5.$$

Therefore,
$$\sigma_{\widehat{\alpha}_i} = \sigma_{\widehat{\alpha}_j} = \sqrt{2}, \quad \tau_i = \tau_j = \sqrt{2}, \quad \delta_i = \delta_j = 1.$$

The Monte Carlo results are obtained by the following steps:

(S1) We independently generate 10^5 random numbers [5] for
$$\widehat{\alpha}_i \stackrel{\text{ind.}}{\sim} N\left(\alpha_i, \frac{\sigma^2}{\lambda_i}\right), \quad \widehat{\alpha}_j \stackrel{\text{ind.}}{\sim} N\left(\alpha_j, \frac{\sigma^2}{\lambda_j}\right), \quad \text{and} \quad \widehat{\sigma}^2 \sim \sigma^2 \chi_\nu^2,$$
respectively.

(S2) Using these random numbers in Step (S1), two sets of 10^5 feasible GRR estimates are given by
$$\widehat{\alpha}_i^* = \frac{\lambda_i}{\lambda_i + \widehat{k}_i^*}\widehat{\alpha}_i = \frac{\lambda_i}{\lambda_i + \widehat{\sigma}^2/\widehat{\alpha}_i^2}\widehat{\alpha}_i \quad \text{and} \quad \widehat{\alpha}_j^* = \frac{\lambda_j}{\lambda_j + \widehat{k}_j^*}\widehat{\alpha}_j = \frac{\lambda_j}{\lambda_j + \widehat{\sigma}^2/\widehat{\alpha}_j^2}\widehat{\alpha}_j.$$

(S3) Finally, we calculate the target sample moments from these sets.

In Table 5.3, there are differences from the fourth decimal place between the exact and the Monte Carlo cases in the first and second moments, and from the second decimal place in the cross moment. Note that that all digits are significant in the results for the exact case.

---
[5] We use physical random numbers generated by the random number generator "Random Streamer RPG102" produced by Nippon Techno Lab., Inc.

5.5.4 Relative Efficiency

Let us consider several numerical evaluations of the relative efficiency (5.44) and (5.47):

$$\text{REff}(\widehat{\alpha}_i^*, \widehat{\alpha}_i) = \begin{cases} 2\delta_i(\mu_{2i} - 2\mu_{1i} + 1) & \text{if } \delta_i \neq 0, \\ \widetilde{\mu}_2(\nu) & \text{if } \delta_i = 0. \end{cases}$$

Note that such evaluations essentially depend on the terms μ_{1i} and μ_{2i} defined by (5.20) and (5.21), respectively. We need to take $(L, M) = (900, 100)$ for enough accuracy with respect to convergence of $\mu_1(\delta_i, \nu, L, M)$ and $\mu_2(\delta_i, \nu, L, M)$ for the following values:

$$\begin{cases} \delta_i = 0, 0.01, 0.05, 0.1, 0.2, 0.5, 0.7, 0.9, 1, 2, 5, 10, 20, 50, \\ \nu = 1, 2, 5, 10, 15, 20, 30, 40, 50, 100. \end{cases} \quad (5.51)$$

Numerical results for relative efficiency for the cases of (5.51) are listed in Table 5.4

Remark 5.6. *Relative efficiencies for the same values of (δ_i, ν) were also given in Dwivedi et al. (1980), Hemmerle and Carey (1983), and Jimichi (2005). However, those results seem to be lacking in sufficient accuracy due to limitations in the available computational environment.*

Remark 5.7. *From Table 5.4, we know that*

$$REff(\widehat{\alpha}_i^*, \widehat{\alpha}_i) < 1 \iff MSE(\widehat{\alpha}_i^*) < MSE(\widehat{\alpha}_i)$$

for $\nu \geq 2$, $\delta_i \leq 1 (\Leftrightarrow \tau_i^2 \leq 2)$. (See also Figure 5.6.) These results suggest the following method for choosing k_i:

$$k_i = \begin{cases} \widehat{k}_i^*, & \text{if } \delta_i \leq 1, \\ 0, & \text{if } \delta_i > 1. \end{cases} \quad (5.52)$$

Therefore, the following algorithm for choosing k_i is proposed for when $\nu \geq 3$:

$$\widetilde{k}_i^* = \begin{cases} \widehat{k}_i^*, & \text{if } \widehat{\delta}_i \leq 1, \\ 0, & \text{if } \widehat{\delta}_i > 1, \end{cases} \quad (5.53)$$

where

$$\widehat{\delta}_i := \frac{1}{2} \widehat{\tau}_i^2 := \frac{1}{2} \frac{\widehat{\alpha}_i^2}{\widehat{\sigma}^2 / \lambda_i}. \quad (5.54)$$

Note that this algorithm is an improvement on the version of the $\widehat{k}_i^*$ algorithm in (4.5) made by using the results with respect to the exact moments of the feasible GRR estimator $\widehat{\alpha}_i^*$. See also Section 5 in Dwivedi et al. (1980). Additional algorithms for choosing k_i are given in Hemmerle and Carey (1983) and Firinguetti (1999).

5.5. NUMERICAL EVALUATIONS

Table 5.4: Relative efficiency ($\times 100$) of the OLS relative to the feasible GRR estimator: $(L, M) = (900, 100)$

$\delta_i \backslash \nu$	1	2	5	10	15	20	30	40	50	100
0	62.500	56.194	50.975	48.919	48.198	47.831	47.459	47.272	47.159	46.932
0.01	63.060	56.847	51.701	49.674	48.962	48.600	48.233	48.048	47.937	47.713
0.05	65.253	59.398	54.542	52.626	51.952	51.609	51.262	51.086	50.981	50.769
0.10	67.892	62.465	57.955	56.171	55.542	55.222	54.898	54.735	54.636	54.438
0.20	72.843	68.209	64.341	62.802	62.259	61.982	61.701	61.559	61.473	61.301
0.50	85.388	82.690	80.393	79.460	79.127	78.955	78.781	78.693	78.640	78.533
0.70	92.120	90.404	88.907	88.283	88.058	87.941	87.823	87.762	87.726	87.652
0.90	97.778	96.844	95.984	95.609	95.471	95.398	95.324	95.286	95.263	95.217
1.00	100.257	99.649	99.056	98.785	98.683	98.630	98.574	98.546	98.529	98.493
2.00	115.883	116.927	117.683	117.947	118.034	118.077	118.120	118.141	118.154	118.179
5.00	123.249	123.500	123.612	123.636	123.642	123.644	123.646	123.646	123.647	123.647
10.00	117.501	116.323	115.391	115.028	114.901	114.835	114.769	114.736	114.716	114.675
20.00	110.531	109.169	108.218	107.873	107.755	107.695	107.635	107.605	107.587	107.550
50.00	104.676	103.875	103.363	103.187	103.127	103.098	103.068	103.053	103.044	103.026

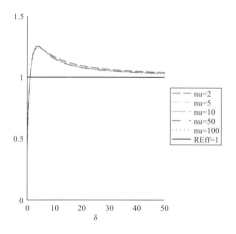

Figure 5.5: Efficiency of the OLS estimator $\widehat{\alpha}_i$ relative to the feasible GRR estimator $\widehat{\alpha}_i^*$: $0 \leq \delta_i \leq 50$

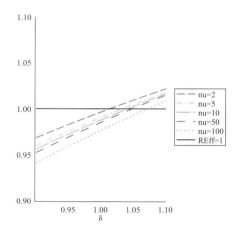

Figure 5.6: Efficiency of the OLS estimator $\widehat{\alpha}_i$ relative to the feasible GRR estimator $\widehat{\alpha}_i^*$: $0.9 \leq \delta_i \leq 1.1$

Chapter 6
Estimation of MSE Criteria by Bootstrap Method

In this chapter, we estimate the MSE criteria of the shrinkage(-type) regression estimators by the parametric bootstrap method, and evaluate their improvements by comparing the results of several Monte Carlo simulations under multicollinearity.

6.1 Introduction

The *bootstrap* [1] method was introduced by Efron (1979) and is one of a larger class of methods that resample from an original data set. See Efron and Tibshirani (1993) or Davison and Hinkley (1997) for a detailed explanation of the several bootstrap methods.

Traditional methods used for evaluating improvement in accuracy of a shrinkage regression estimator have depended on the unknown true values of the parameters. For example, the feasible ORR estimator $\widehat{\boldsymbol{\beta}}_p(\widehat{k})$ often has been evaluated by the *total average squared error* (TASE)

$$\text{TASE}(\widehat{\boldsymbol{\beta}}_p(\widehat{k})) := \frac{1}{N} \sum_{j=1}^{N} \|\widehat{\boldsymbol{\beta}}_p(\widehat{k})_{(j)} - \boldsymbol{\beta}_p\|^2,$$

where $\widehat{\boldsymbol{\beta}}_p(\widehat{k})_{(j)}$ is the feasible ORR estimate obtained by the jth (Monte Carlo) simulation, and N is the number of iterations. However, this evaluation method could not be actually used because the parameter vector $\boldsymbol{\beta}_p$ is unknown. This is because we usually have only the data set in practice, whereas the criterion needs the true values of the parameters. To address this problem, we discuss several estimations of the MSE criteria of the regression estimators by the *parametric bootstrap method*. Note that the procedures in the method are just "algorithms" for estimating the MSE criteria, and we can objectively evaluate the improvements of the accuracy of the feasible shrinkage regression estimators with only the data set.

---

[1] Efron and Tibshirani (1993) described the origin of the term *bootstrap* as follows:

> The use of the term bootstrap derives from the phrase *to pull oneself up by one's bootstrap, widely thought to be based on one of the eighteenth century Adventures of Baron Munchausen, by Rudolph Eric Raspe.*

See p. 5 in Efron and Tibshirani (1993) for details.

6.2 Parametric Bootstrap Method in Regression Analysis

Assume that the distribution of the errors is normal:

$$\varepsilon_i \overset{\text{i.i.d.}}{\sim} N(0, \sigma^2), \quad i = 1, \cdots, n.$$

Therefore, the models (M_Z), (M_X), and (M_A) are the normal linear models (Gauss-Markov setups with normality) $\text{GMN}(\boldsymbol{Y}, \mathbf{Z}\boldsymbol{\theta}, \sigma^2 \mathbf{I}_n)$, $\text{GMN}(\boldsymbol{Y}, \mathbf{X}\boldsymbol{\beta}, \sigma^2 \mathbf{I}_n)$, and $\text{GMN}(\boldsymbol{Y}, \mathbf{A}\boldsymbol{\alpha}, \sigma^2 \mathbf{I}_n)$, respectively.

Under the models, the vector of the response variable $\boldsymbol{Y}$ is normally distributed. For example,

$$\boldsymbol{Y} \sim N(\mathbf{X}\boldsymbol{\beta}, \sigma^2 \mathbf{I}_n) =: G(\boldsymbol{y} \, ; \, \boldsymbol{\eta}_\beta); \text{ under the standardized model } (M_X),$$

where $\boldsymbol{\eta}_\beta$ is defined by

$$\boldsymbol{\eta}_\beta := \begin{bmatrix} \boldsymbol{\beta} \\ \sigma^2 \end{bmatrix}.$$

The vector of estimates for $\boldsymbol{\eta}_\beta$ is given by

$$\widehat{\boldsymbol{\eta}}_\beta := \begin{bmatrix} \widehat{\boldsymbol{\beta}} \\ \widehat{\sigma}^2 \end{bmatrix},$$

where $\widehat{\boldsymbol{\beta}}$ is the OLS estimate defined by

$$\widehat{\boldsymbol{\beta}} = (\mathbf{X}'\mathbf{X})^{-1} \mathbf{X}'\boldsymbol{y},$$

with $\boldsymbol{y}$ being an observed vector of $\boldsymbol{Y}$, and

$$\widehat{\sigma}^2 = \frac{1}{n-p-1}(\boldsymbol{y} - \mathbf{X}\widehat{\boldsymbol{\beta}})'(\boldsymbol{y} - \mathbf{X}\widehat{\boldsymbol{\beta}})$$

is an estimate for σ^2. The unknown distribution $G_\beta := G(\boldsymbol{y} \, ; \, \boldsymbol{\eta}_\beta)$ of the vector of the response variables $\boldsymbol{Y}$ in the model (M_X) is estimated by $\widehat{G}_\beta := G(\boldsymbol{y} \, ; \, \widehat{\boldsymbol{\eta}}_\beta)$:

$$G_\beta := G(\boldsymbol{y} \, ; \, \boldsymbol{\eta}_\beta) \xleftarrow{\text{estimate}} \widehat{G}_\beta := G(\boldsymbol{y} \, ; \, \widehat{\boldsymbol{\eta}}_\beta).$$

The estimating method based on the random sample

$$\boldsymbol{Y}^B \sim \widehat{G}_\beta = G(\boldsymbol{y} \, ; \, \widehat{\boldsymbol{\eta}}_\beta) = N(\mathbf{X}\widehat{\boldsymbol{\beta}}, \widehat{\sigma}^2 \mathbf{I}_n)$$

is called the *parametric bootstrap method* under the normal linear model $\text{GMN}(\boldsymbol{Y}, \mathbf{X}\boldsymbol{\beta}, \sigma^2 \mathbf{I}_n)$.

6.3 Parametric Bootstrap Method for OLS Estimator

The OLS estimator by the parametric bootstrap method for $\boldsymbol{\beta}_p$ in the model (M_X) is defined by

$$\widehat{\boldsymbol{\beta}}_p^B := (\mathbf{X}_p'\mathbf{X}_p)^{-1}\mathbf{X}_p'\boldsymbol{Y}^B \tag{6.1}$$

and is called the *parametric bootstrap OLS estimator* for $\boldsymbol{\beta}_p$. Furthermore, the parametric bootstrap OLS estimator for $\boldsymbol{\theta}$ in the model (M_Z) is defined by

$$\widehat{\boldsymbol{\theta}}^B := (\mathbf{Z}'\mathbf{Z})^{-1}\mathbf{Z}'\boldsymbol{Y}^B. \tag{6.2}$$

The bias vector, the variance-covariance matrix, the MSE matrix, and the TMSE of the OLS estimators $\widehat{\boldsymbol{\beta}}_p$, $\widehat{\boldsymbol{\theta}}$ under the distribution $G := G_\beta$ are equivalent to the results under the distribution F given in Subsection 3.2.1. For example,

$$\mathrm{MSE}_G(\widehat{\boldsymbol{\beta}}_p) = \mathrm{V}_G(\widehat{\boldsymbol{\beta}}_p) = \sigma^2(\mathbf{X}_p'\mathbf{X}_p)^{-1},$$

$$\mathrm{TMSE}_G(\widehat{\boldsymbol{\beta}}_p) = trace\, \mathrm{MSE}_G(\widehat{\boldsymbol{\beta}}_p) = \sigma^2 \sum_{i=1}^{p} \frac{1}{\lambda_i},$$

$$\mathrm{MSE}_G(\widehat{\boldsymbol{\theta}}) = \mathrm{V}_G(\widehat{\boldsymbol{\theta}}) = \sigma^2\, (\mathbf{Z}'\mathbf{Z})^{-1},$$

$$\mathrm{TMSE}_G(\widehat{\boldsymbol{\theta}}) = trace\, \mathrm{MSE}_G(\widehat{\boldsymbol{\theta}}) = \sigma^2 trace\, (\mathbf{Z}'\mathbf{Z})^{-1}.$$

The MSE matrix and the TMSE of the parametric bootstrap OLS estimators $\widehat{\boldsymbol{\beta}}_p^B$, $\widehat{\boldsymbol{\theta}}^B$ under the distribution $\widehat{G} := \widehat{G}_\beta$ can be obtained as follows:

$$\mathrm{MSE}_{\widehat{G}}(\widehat{\boldsymbol{\beta}}_p^B) = \mathrm{V}_{\widehat{G}}(\widehat{\boldsymbol{\beta}}_p^B) = \widehat{\sigma}^2(\mathbf{X}_p'\mathbf{X}_p)^{-1},$$

$$\mathrm{TMSE}_{\widehat{G}}(\widehat{\boldsymbol{\beta}}_p^B) = trace\, \mathrm{MSE}_{\widehat{G}}(\widehat{\boldsymbol{\beta}}_p^B) = \widehat{\sigma}^2 \sum_{i=1}^{p} \frac{1}{\lambda_i},$$

$$\mathrm{MSE}_{\widehat{G}}(\widehat{\boldsymbol{\theta}}^B) = \mathrm{V}_{\widehat{G}}(\widehat{\boldsymbol{\theta}}^B) = \widehat{\sigma}^2\, (\mathbf{Z}'\mathbf{Z})^{-1},$$

$$\mathrm{TMSE}_{\widehat{G}}(\widehat{\boldsymbol{\theta}}^B) = trace\, \mathrm{MSE}_{\widehat{G}}(\widehat{\boldsymbol{\theta}}^B) = \widehat{\sigma}^2 trace\, (\mathbf{Z}'\mathbf{Z})^{-1}.$$

Note that $\mathrm{MSE}_{\widehat{G}}(\cdot)$ is called the *bootstrap estimate* for $\mathrm{MSE}_G(\cdot)$, and $\mathrm{TMSE}_{\widehat{G}}(\cdot)$ is that for $\mathrm{TMSE}_G(\cdot)$. However, it is generally difficult to obtain bootstrap estimates for them (especially, for the shrinkage regression estimators), and if we want to estimate them by the bootstrap method, then we usually use the following *Monte Carlo approximations*:

$$\widehat{\mathrm{MSE}}_G(\widehat{\boldsymbol{\beta}}_p) := \widehat{\mathrm{V}}_G(\widehat{\boldsymbol{\beta}}_p) + \widehat{\mathrm{bias}}_G(\widehat{\boldsymbol{\beta}}_p)\widehat{\mathrm{bias}}_G(\widehat{\boldsymbol{\beta}}_p)',$$

$$\widehat{\mathrm{TMSE}}_G(\widehat{\boldsymbol{\beta}}_p) := trace\widehat{\mathrm{MSE}}_G(\widehat{\boldsymbol{\beta}}_p),$$

$$\widehat{\mathrm{MSE}}_G(\widehat{\boldsymbol{\theta}}) := \widehat{\mathrm{V}}_G(\widehat{\boldsymbol{\theta}}) + \widehat{\mathrm{bias}}_G(\widehat{\boldsymbol{\theta}})\widehat{\mathrm{bias}}_G(\widehat{\boldsymbol{\theta}})',$$

$$\widehat{\mathrm{TMSE}}_G(\widehat{\boldsymbol{\theta}}) := trace\widehat{\mathrm{MSE}}_G(\widehat{\boldsymbol{\theta}}),$$

where

$$\widehat{\text{bias}}_G(\widehat{\boldsymbol{\beta}}_p) := \frac{1}{B}\sum_{b=1}^{B}\widehat{\boldsymbol{\beta}}_{p(b)} - \widehat{\boldsymbol{\beta}}_p =: \overline{\widehat{\boldsymbol{\beta}}}_p - \widehat{\boldsymbol{\beta}}_p,$$

$$\widehat{\text{V}}_G(\widehat{\boldsymbol{\beta}}_p) := \frac{1}{B-1}\sum_{b=1}^{B}\left(\widehat{\boldsymbol{\beta}}_{p(b)} - \overline{\widehat{\boldsymbol{\beta}}}_p\right)\left(\widehat{\boldsymbol{\beta}}_{p(b)} - \overline{\widehat{\boldsymbol{\beta}}}_p\right)',$$

$$\widehat{\text{bias}}_G(\widehat{\boldsymbol{\theta}}) := \frac{1}{B}\sum_{b=1}^{B}\widehat{\boldsymbol{\theta}}_{(b)} - \widehat{\boldsymbol{\theta}} =: \overline{\widehat{\boldsymbol{\theta}}} - \widehat{\boldsymbol{\theta}},$$

$$\widehat{\text{V}}_G(\widehat{\boldsymbol{\theta}}) := \frac{1}{B-1}\sum_{b=1}^{B}\left(\widehat{\boldsymbol{\theta}}_{(b)} - \overline{\widehat{\boldsymbol{\theta}}}\right)\left(\widehat{\boldsymbol{\theta}}_{(b)} - \overline{\widehat{\boldsymbol{\theta}}}\right)',$$

and $\boldsymbol{y}_{(b)}$ ($b = 1, \cdots, B$) is the bth generated realization of $\boldsymbol{Y}^B \sim \widehat{G} = N(\mathbf{X}\widehat{\boldsymbol{\beta}}, \widehat{\sigma}^2 \mathbf{I}_n)$ and

$$\widehat{\boldsymbol{\beta}}_{(b)} := (\mathbf{X}'\mathbf{X})^{-1}\mathbf{X}'\boldsymbol{y}_{(b)} = \begin{bmatrix} \overline{y}_{(b)} \\ (\mathbf{X}'_p\mathbf{X}_p)^{-1}\mathbf{X}'_p\boldsymbol{y}_{(b)} \end{bmatrix} =: \begin{bmatrix} \widehat{\beta}_{0(b)} \\ \widehat{\boldsymbol{\beta}}_{p(b)} \end{bmatrix},$$

$$\widehat{\boldsymbol{\theta}}_{(b)} := \mathbf{T}^{-1}\widehat{\boldsymbol{\beta}}_{(b)} = (\mathbf{Z}'\mathbf{Z})^{-1}\mathbf{Z}'\boldsymbol{y}_{(b)}.$$

6.4 Parametric Bootstrap Method for Feasible Shrinkage Regression Estimators

Let $\widetilde{\boldsymbol{\beta}}_p$ and $\widetilde{\boldsymbol{\theta}}$ be the feasible shrinkage regression estimator for $\boldsymbol{\beta}_p$ and $\boldsymbol{\theta}$, respectively. (See also Chapter 4.) Additionally, their *parametric bootstrap shrinkage regression estimators* are denoted by $\widetilde{\boldsymbol{\beta}}_p^B$, $\widetilde{\boldsymbol{\theta}}^B$. Although it is difficult to obtain the bootstrap estimate $\text{MSE}_{\widehat{G}}(\widetilde{\boldsymbol{\beta}}_p^B)$, $\text{MSE}_{\widehat{G}}(\widetilde{\boldsymbol{\theta}}^B)$ for the MSE matrix of the feasible shrinkage regression estimators under the distribution $\widehat{G}$ because it is hard to calculate their exact moments under the distribution G, we can estimate the MSE matrix and the TMSE of the shrinkage regression estimators by Monte Carlo approximations from an analogy of the case of the OLS estimator. That is,

$$\widehat{\text{MSE}}_G(\widetilde{\boldsymbol{\beta}}_p) := \widehat{\text{V}}_G(\widetilde{\boldsymbol{\beta}}_p) + \widehat{\text{bias}}_G(\widetilde{\boldsymbol{\beta}}_p)\widehat{\text{bias}}_G(\widetilde{\boldsymbol{\beta}}_p)',$$
$$\widehat{\text{TMSE}}_G(\widetilde{\boldsymbol{\beta}}_p) := trace\widehat{\text{MSE}}_G(\widetilde{\boldsymbol{\beta}}_p),$$
$$\widehat{\text{MSE}}_G(\widetilde{\boldsymbol{\theta}}) := \widehat{\text{V}}_G(\widetilde{\boldsymbol{\theta}}) + \widehat{\text{bias}}_G(\widetilde{\boldsymbol{\theta}})\widehat{\text{bias}}_G(\widetilde{\boldsymbol{\theta}})',$$
$$\widehat{\text{TMSE}}_G(\widetilde{\boldsymbol{\theta}}) := trace\widehat{\text{MSE}}_G(\widetilde{\boldsymbol{\theta}}),$$

6.4. PARAMETRIC BOOTSTRAP METHOD FOR FEASIBLE SR ESTIMATORS

where

$$\widehat{\text{bias}}_G(\widetilde{\boldsymbol{\beta}}_p) := \frac{1}{B}\sum_{b=1}^{B}\widetilde{\boldsymbol{\beta}}_{p(b)} - \widehat{\boldsymbol{\beta}}_p =: \overline{\widetilde{\boldsymbol{\beta}}}_p - \widehat{\boldsymbol{\beta}}_p,$$

$$\widehat{\text{V}}_G(\widetilde{\boldsymbol{\beta}}_p) := \frac{1}{B-1}\sum_{b=1}^{B}\left(\widetilde{\boldsymbol{\beta}}_{p(b)} - \overline{\widetilde{\boldsymbol{\beta}}}_p\right)\left(\widetilde{\boldsymbol{\beta}}_{p(b)} - \overline{\widetilde{\boldsymbol{\beta}}}_p\right)',$$

$$\widehat{\text{bias}}_G(\widetilde{\boldsymbol{\theta}}) := \frac{1}{B}\sum_{b=1}^{B}\widetilde{\boldsymbol{\theta}}_{(b)} - \widehat{\boldsymbol{\theta}} =: \overline{\widetilde{\boldsymbol{\theta}}} - \widehat{\boldsymbol{\theta}},$$

$$\widehat{\text{V}}_G(\widetilde{\boldsymbol{\theta}}) := \frac{1}{B-1}\sum_{b=1}^{B}\left(\widetilde{\boldsymbol{\theta}}_{(b)} - \overline{\widetilde{\boldsymbol{\theta}}}\right)\left(\widetilde{\boldsymbol{\theta}}_{(b)} - \overline{\widetilde{\boldsymbol{\theta}}}\right)',$$

and $\widetilde{\boldsymbol{\beta}}_{p(b)}$, $\widetilde{\boldsymbol{\theta}}_{(b)}$ constitute the feasible shrinkage regression estimate based on $\boldsymbol{y}_{(b)}$.

Remark 6.1. Since the exact moments of the feasible GRR estimators $\widehat{\boldsymbol{\beta}}_p^* := \widehat{\boldsymbol{\beta}}_p(\widehat{\mathbf{K}}_p^*)$ are known from the results in Chapter 5, we can use the bootstrap estimate $MSE_{\widehat{G}}(\widehat{\boldsymbol{\beta}}_p^{*B})$. That is, from $\widehat{\boldsymbol{\beta}}_p^* = \boldsymbol{\Gamma}_p\widehat{\boldsymbol{\alpha}}_p^*$ and the MSE matrix of the feasible GRR estimator $\widehat{\boldsymbol{\alpha}}_p^*$ (see (5.36)), the MSE matrix of the feasible GRR estimator $\widehat{\boldsymbol{\beta}}_p^*$ is obtained as follows:

$$MSE_G(\widehat{\boldsymbol{\beta}}_p^*) = \boldsymbol{\Gamma}_p MSE_G(\widehat{\boldsymbol{\alpha}}_p^*)\boldsymbol{\Gamma}_p'$$
$$= \boldsymbol{\Gamma}_p\left(V_G(\widehat{\boldsymbol{\alpha}}_p^*) + bias_G(\widehat{\boldsymbol{\alpha}}_p^*)bias_G(\widehat{\boldsymbol{\alpha}}_p^*)'\right)\boldsymbol{\Gamma}_p'$$
$$= \boldsymbol{\Gamma}_p\left(\mathcal{A}(\mathbf{M}_2 - \boldsymbol{\mu}_1\boldsymbol{\mu}_1')\mathcal{A} + (\mathbf{M}_1 - \mathbf{I}_p)\boldsymbol{\alpha}_p\boldsymbol{\alpha}_p'(\mathbf{M}_1 - \mathbf{I}_p)\right)\boldsymbol{\Gamma}_p', \quad (6.3)$$

where $\mathcal{A} = diag(\alpha_1,\ldots,\alpha_p)$ and

$$\boldsymbol{\mu}_1 = \begin{bmatrix} \mu_{11} \\ \vdots \\ \mu_{1p} \end{bmatrix}, \quad \mathbf{M}_1 = \begin{bmatrix} \mu_{11} & & \mathbf{O} \\ & \ddots & \\ \mathbf{O} & & \mu_{1p} \end{bmatrix}, \quad \mathbf{M}_2 = \begin{bmatrix} \mu_{21} & \cdots & \mu_{111p} \\ \vdots & \ddots & \vdots \\ \mu_{11p1} & \cdots & \mu_{2p} \end{bmatrix},$$

and recall that

$$\mu_{1i} = \mu_1(\delta_i,\nu) = \sum_{\ell=0}^{\infty}\left(\frac{\nu-1}{\nu}\right)^{\ell}\sum_{m=0}^{\infty}\frac{B\left(\ell+\frac{\nu}{2},m+\frac{5}{2}\right)}{B\left(\frac{\nu}{2},m+\frac{3}{2}\right)}p(m;\delta_i),$$

$$\mu_{2i} = \mu_2(\delta_i,\nu) = \frac{1}{2\delta_i}\sum_{\ell=0}^{\infty}(\ell+1)\left(\frac{\nu-1}{\nu}\right)^{\ell}\sum_{m=0}^{\infty}(2m+1)\frac{B\left(\ell+\frac{\nu}{2},m+\frac{7}{2}\right)}{B\left(\frac{\nu}{2},m+\frac{3}{2}\right)}p(m;\delta_i),$$

$$\mu_{11ij} := \mu_{11}(\delta_i,\delta_j,\nu) = \sum_{\ell_1=0}^{\infty}\sum_{\ell_2=0}^{\infty}\left(\frac{\nu-1}{\nu}\right)^{\ell_1+\ell_2}$$
$$\times \sum_{m_1=0}^{\infty}\sum_{m_2=0}^{\infty}\frac{\mathcal{B}_{\ell_1+1\ \ell_2+1}\left(m_1+\frac{5}{2},m_2+\frac{5}{2},\ell_1+\ell_2+\frac{\nu}{2}\right)}{B\left(m_1+\frac{3}{2},m_2+\frac{3}{2},\frac{\nu}{2}\right)}p(m_1;\delta_i)p(m_2;\delta_j).$$

Therefore, the bootstrap estimate $MSE_{\widehat{G}}(\widehat{\boldsymbol{\beta}}_p^{*B})$ is

$$\begin{aligned}
MSE_{\widehat{G}}(\widehat{\boldsymbol{\beta}}_p^{*B}) &= \boldsymbol{\Gamma}_p MSE_{\widehat{G}}(\widehat{\boldsymbol{\alpha}}_p^{*B})\boldsymbol{\Gamma}_p' \\
&= \boldsymbol{\Gamma}_p \left(V_{\widehat{G}}(\widehat{\boldsymbol{\alpha}}_p^{*B}) + bias_{\widehat{G}}(\widehat{\boldsymbol{\alpha}}_p^{*B})bias_{\widehat{G}}(\widehat{\boldsymbol{\alpha}}_p^{*B})' \right) \boldsymbol{\Gamma}_p' \\
&= \boldsymbol{\Gamma}_p \left(\widehat{\mathcal{A}}(\widehat{\mathbf{M}}_2 - \widehat{\boldsymbol{\mu}}_1\widehat{\boldsymbol{\mu}}_1')\widehat{\mathcal{A}} + (\widehat{\mathbf{M}}_1 - \mathbf{I}_p)\widehat{\boldsymbol{\alpha}}_p\widehat{\boldsymbol{\alpha}}_p'(\widehat{\mathbf{M}}_1 - \mathbf{I}_p) \right) \boldsymbol{\Gamma}_p',
\end{aligned} \tag{6.4}$$

where $\widehat{\boldsymbol{\alpha}}_p = [\widehat{\alpha}_1, \ldots, \widehat{\alpha}_p]'$ is the OLS estimate for $\boldsymbol{\alpha}_p$ and $\widehat{\mathcal{A}} := diag(\widehat{\alpha}_1, \ldots, \widehat{\alpha}_p)$,

$$\widehat{\boldsymbol{\mu}}_1 := \begin{bmatrix} \widehat{\mu}_{11} \\ \vdots \\ \widehat{\mu}_{1p} \end{bmatrix}, \quad \widehat{\mathbf{M}}_1 := \begin{bmatrix} \widehat{\mu}_{11} & & \mathbf{O} \\ & \ddots & \\ \mathbf{O} & & \widehat{\mu}_{1p} \end{bmatrix}, \quad \widehat{\mathbf{M}}_2 := \begin{bmatrix} \widehat{\mu}_{21} & \cdots & \widehat{\mu}_{111p} \\ \vdots & \ddots & \vdots \\ \widehat{\mu}_{11p1} & \cdots & \widehat{\mu}_{2p} \end{bmatrix}.$$

$\widehat{\mu}_{1i} := \mu_1(\widehat{\delta}_i, \nu)$, $\widehat{\mu}_{2i} := \mu_2(\widehat{\delta}_i, \nu)$, $\widehat{\mu}_{11ij} := \mu_{11}(\widehat{\delta}_i, \widehat{\delta}_j, \nu)$, where

$$\widehat{\delta}_i := \frac{1}{2}\widehat{\tau}_i^2 := \frac{1}{2}\frac{\widehat{\alpha}_i^2}{\widehat{\sigma}^2/\lambda_i}$$

and $\widehat{\sigma}^2$ is the estimate σ^2.

6.5 Monte Carlo Simulations

In this section, we calculate some statistics, including the TMSEs of the feasible shrinkage regression estimators. Then, we use two data sets with multicollinearity given by Hoerl (1962) (see Appendix C) and evaluate the estimations of the TMSEs of the estimators by carrying out Monte Carlo simulations based on the parametric bootstrap method [2]. Note that the number of bootstrap iteration is $B = 10^5$, which is sufficient for our simulations. (See Figures 6.1 and 6.2.)

The results of the simulations are given in Tables 6.1, 6.2, 6.3, 6.4. Let us explain the notation used in these tables. "$\widetilde{\text{TMSE}}_G$" is the Monte Carlo approximation of the TMSE of the regression estimators, "TMSE_G" is the exact TMSE, "$\text{TMSE}_{\widehat{G}}$" is the bootstrap estimate for the TMSE of the regression estimators, and "TMSE*" is the minimum of the TMSE for the (usual) shrinkage regression estimators where the number of principal components r is chosen by $r^*(=1)$ (3.39) of the PCR estimator (see Appendix C), and $k_i^* = \sigma^2/\alpha_i^2$ of the GRR estimator. The "optimum" ridge coefficient k^* for the ORR estimator and the r-k class estimator are chosen by Newton's method. Furthermore, the ridge coefficients k_0 in the ORR, GRR, and r-k-class-type estimators for $\boldsymbol{\theta}$ are chosen according to (3.51), (3.52), and (3.53),

---

[2] We use the random number generator "Random Streamer RPG102" produced by Nippon Techno Lab., Inc. to generate the random numbers used in our simulations.

respectively. Several representative graphs of the TMSEs of the shrinkage regression estimators for $\boldsymbol{\beta}_p$ are given in Figures 6.3, 6.4, and for $\boldsymbol{\theta}$ and in Figures 6.5 and 6.6, where the following values of k are used:

$$k = 0,\ 0.0001,\ 0.001(0.001)0.02,\ 0.03(0.01)0.1,\ 0.11(0.01)0.5,\ 0.6(0.05)1.0.$$

For example, these statistics in the case of the feasible GRR estimator $\widehat{\boldsymbol{\beta}}_p^*$ are as follows:

$$\widehat{\text{TMSE}}_G(\widehat{\boldsymbol{\beta}}_p^*) = trace\widehat{\text{MSE}}_G(\widehat{\boldsymbol{\beta}}_p^*) = trace\widehat{\text{V}}_G(\widehat{\boldsymbol{\beta}}_p) + \widehat{\text{bias}}_G(\widehat{\boldsymbol{\beta}}_p)'\widehat{\text{bias}}_G(\widehat{\boldsymbol{\beta}}_p), \quad (6.5)$$

where

$$\widehat{\text{bias}}_G(\widehat{\boldsymbol{\beta}}_p^*) := \frac{1}{B}\sum_{b=1}^{B} \widehat{\boldsymbol{\beta}}_{p(b)}^* - \widehat{\boldsymbol{\beta}}_p =: \overline{\widehat{\boldsymbol{\beta}}}_p^* - \widehat{\boldsymbol{\beta}}_p,$$

$$\widehat{\text{V}}_G(\widehat{\boldsymbol{\beta}}_p^*) := \frac{1}{B-1}\sum_{b=1}^{B} \left(\widehat{\boldsymbol{\beta}}_{p(b)}^* - \overline{\widehat{\boldsymbol{\beta}}}_p^*\right)\left(\widehat{\boldsymbol{\beta}}_{p(b)}^* - \overline{\widehat{\boldsymbol{\beta}}}_p^*\right)'.$$

$$\begin{aligned}
\text{TMSE}_G(\widehat{\boldsymbol{\beta}}_p^*) &= trace\text{MSE}_G(\widehat{\boldsymbol{\beta}}_p^*) = trace\text{MSE}_G(\widehat{\boldsymbol{\alpha}}_p^*)\\
&= trace\text{V}_G(\widehat{\boldsymbol{\alpha}}_p^*) + \text{bias}_G(\widehat{\boldsymbol{\alpha}}_p^*)'\text{bias}_G(\widehat{\boldsymbol{\alpha}}_p^*)\\
&= trace\mathcal{A}(\mathbf{M}_2 - \boldsymbol{\mu}_1\boldsymbol{\mu}_1')\mathcal{A} + \boldsymbol{\alpha}_p'(\mathbf{M}_1 - \mathbf{I}_p)(\mathbf{M}_1 - \mathbf{I}_p)\boldsymbol{\alpha}_p\\
&= \sum_{i=1}^{p} \alpha_i^2(\mu_{2i} - 2\mu_{1i} + 1),
\end{aligned} \quad (6.6)$$

$$\begin{aligned}
\text{TMSE}_{\widehat{G}}(\widehat{\boldsymbol{\beta}}_p^{*B}) &= trace\text{MSE}_{\widehat{G}}(\widehat{\boldsymbol{\beta}}_p^{*B}) = trace\text{MSE}_{\widehat{G}}(\widehat{\boldsymbol{\alpha}}_p^{*B})\\
&= trace\text{V}_{\widehat{G}}(\widehat{\boldsymbol{\alpha}}_p^{*B}) + \text{bias}_{\widehat{G}}(\widehat{\boldsymbol{\alpha}}_p^{*B})'\text{bias}_{\widehat{G}}(\widehat{\boldsymbol{\alpha}}_p^{*B})\\
&= trace\widehat{\mathcal{A}}(\widehat{\mathbf{M}}_2 - \widehat{\boldsymbol{\mu}}_1\widehat{\boldsymbol{\mu}}_1')\widehat{\mathcal{A}} + \widehat{\boldsymbol{\alpha}}_p'(\widehat{\mathbf{M}}_1 - \mathbf{I}_p)(\widehat{\mathbf{M}}_1 - \mathbf{I}_p)\widehat{\boldsymbol{\alpha}}_p\\
&= \sum_{i=1}^{p} \widehat{\alpha}_i^2(\widehat{\mu}_{2i} - 2\widehat{\mu}_{1i} + 1),
\end{aligned} \quad (6.7)$$

$$\text{TMSE}^*(\widehat{\boldsymbol{\beta}}_p(\mathbf{K}_p)) = \text{TMSE}(\widehat{\boldsymbol{\beta}}_p(\mathbf{K}_p^*)) (= \min_{\substack{k_i > 0 \\ i=1,\ldots,p}} \text{TMSE}(\widehat{\boldsymbol{\beta}}_p(\mathbf{K}_p)))$$

$$= \sigma^2 \sum_{i=1}^{p} \frac{\lambda_i}{(\lambda_i + k_i^*)^2} + \sum_{i=1}^{p} \frac{k_i^{*2}\alpha_i^2}{(\lambda_i + k_i^*)^2}. \quad (6.8)$$

Note that it is generally difficult to calculate the exact TMSE (TMSE_G) and the bootstrap estimate $\text{TMSE}_{\widehat{G}}$ for the feasible shrinkage regression estimators except in the cases of GRR and OLS.

Let us consider some results of these simulations:

(R1) All of the shrinkage regression estimators are better than the OLS estimator for both data sets.

(R2) The estimates for the TMSEs of the OLS estimators have good accuracy, as is also true in the cases of the feasible GRR estimator.

(R3) Although the exact TMSEs of the shrinkage regression estimators except the OLS and the feasible GRR estimator are not available, the Monte Carlo approximations may be expected to accurately estimate the TMSEs of the feasible shrinkage regression estimators by using the OLS and GRR cases as "benchmarks".

(R4) Although the Monte Carlo approximations have a tendency to underestimate the exact TMSE, the feasible ORR(-type) and the r-k class (-type) estimators have good performances for both data sets.

(R5) The TMSE of the optimal GRR estimator $\widehat{\boldsymbol{\beta}}_p(\mathbf{K}_p^*)$ is the best among these cases, but its feasible version is not good. The reason may depend on the performance of the algorithms for $\widehat{k}_i^*$. Although it is possible to improve them, obtaining the exact moments of the feasible estimator under a new algorithm can be difficult.

Finally, from the above results, the (feasible) shrinkage regression estimators are effective under multicollinearity and their feasibilities could be objectively verified by the parametric bootstrap method. Among these, the feasible ORR estimator may be especially useful for estimating regression coefficients under multicollinearity. Our results here (R5) could be summarized by saying that the shrinkage estimating system strongly depends on the performance of the "plugs".

(Data Set 1)

Table 6.1: Estimated TMSEs of feasible regression estimators for $\boldsymbol{\beta}_p$

Estimators	$\widehat{\mathrm{TMSE}}_G$	TMSE_G	$\mathrm{TMSE}_{\widehat{G}}$	TMSE^*
OLS	34.880	34.518	34.773	34.518
Feasible ORR	6.923	NA	NA	6.415
Feasible GRR	19.850	19.946	19.796	6.001
Feasible PCR	30.817	NA	NA	7.147
Feasible r-k	7.182	NA	NA	7.145

Table 6.2: Estimated TMSEs of feasible regression estimators for $\boldsymbol{\theta}$

Estimators	$\widehat{\mathrm{TMSE}}_G$	TMSE_G	$\mathrm{TMSE}_{\widehat{G}}$	TMSE^*
OLS	19.190	18.984	19.124	18.984
Feasible ORR	5.924	NA	NA	4.888
Feasible GRR	11.854	NA	NA	4.779
Feasible PCR	17.207	NA	NA	5.546
Feasible r-k	5.934	NA	NA	2.940

(Data Set 2)

Table 6.3: Estimated TMSEs of feasible regression estimators for $\boldsymbol{\beta}_p$

Estimators	$\widehat{\mathrm{TMSE}}_G$	TMSE_G	$\mathrm{TMSE}_{\widehat{G}}$	TMSE^*
OLS	83.417	83.688	83.702	83.688
Feasible ORR	2.309	NA	NA	7.079
Feasible GRR	42.132	45.559	42.413	6.598
Feasible PCR	67.550	NA	NA	10.738
Feasible r-k	2.379	NA	NA	10.736

Table 6.4: Estimated TMSEs of feasible regression estimators for $\boldsymbol{\theta}$

Estimators	$\widehat{\mathrm{TMSE}}_G$	TMSE_G	$\mathrm{TMSE}_{\widehat{G}}$	TMSE^*
OLS	57.942	58.110	58.120	58.110
Feasible ORR	3.810	NA	NA	6.376
Feasible GRR	30.449	NA	NA	6.360
Feasible PCR	47.325	NA	NA	8.866
Feasible r-k	3.855	NA	NA	3.477

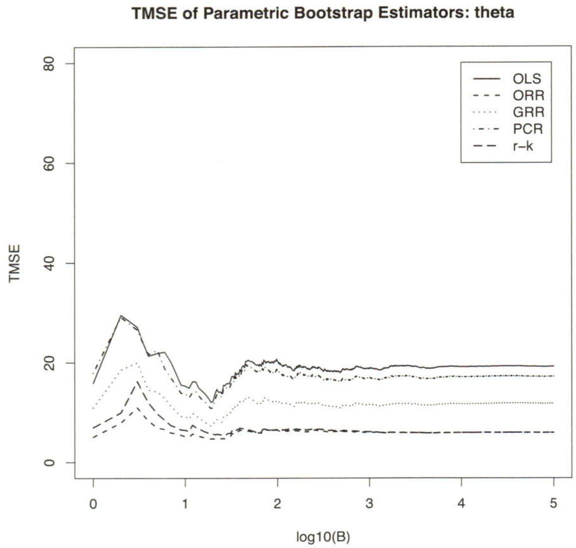

6 ESTIMATION OF MSE CRITERIA BY BOOTSTRAP METHOD

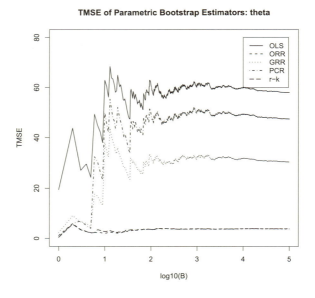

Figure 6.1: Monte Carlo approximations for TMSEs of feasible shrinkage regression estimators for $\boldsymbol{\theta}$ versus (log-scaled) bootstrap iteration number B: Data Set 1

Figure 6.2: Monte Carlo approximations for TMSEs of feasible shrinkage regression estimators for $\boldsymbol{\theta}$ versus (log-scaled) bootstrap iteration number B: Data Set 2

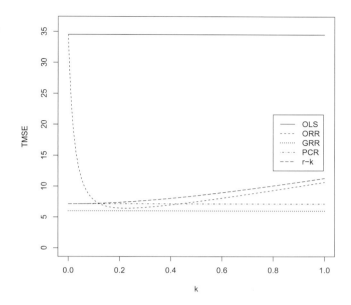

Figure 6.3: TMSEs of OLS $\widehat{\boldsymbol{\beta}}_p$, ORR $\widehat{\boldsymbol{\beta}}_p(\cdot, k)$, GRR $\widehat{\boldsymbol{\beta}}_p(\mathbf{K}_p^*)$, PCR $\widehat{\boldsymbol{\beta}}_p(r^*, \cdot)$, and r-k class $\widehat{\boldsymbol{\beta}}_p(r^*, k)$ estimators versus ridge coefficient k: Data Set 1

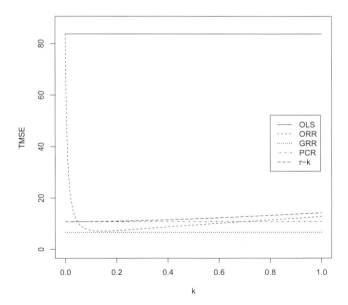

Figure 6.4: TMSEs of OLS $\widehat{\boldsymbol{\beta}}_p$, ORR $\widehat{\boldsymbol{\beta}}_p(\cdot, k)$, GRR $\widehat{\boldsymbol{\beta}}_p(\mathbf{K}_p^*)$, PCR $\widehat{\boldsymbol{\beta}}_p(r^*, \cdot)$, and r-k class $\widehat{\boldsymbol{\beta}}_p(r^*, k)$ estimators versus ridge coefficient k: Data Set 2

84   6 ESTIMATION OF MSE CRITERIA BY BOOTSTRAP METHOD

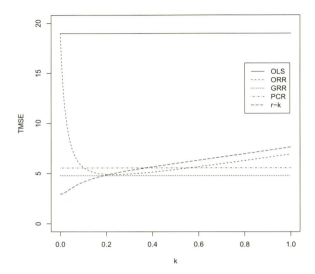

Figure 6.5:   TMSEs of OLS type $\widehat{\boldsymbol{\theta}}$, ORR type $\widehat{\boldsymbol{\theta}}(\cdot, k_0^*)$, GRR type $\widehat{\boldsymbol{\theta}}(k_0^*, \mathbf{K}_p^*)$, PCR type $\widehat{\boldsymbol{\theta}}(r^*, \cdot)$, and r-k class type $\widehat{\boldsymbol{\theta}}(r^*, k_0^*)$ estimators versus ridge coefficient k: Data Set 1

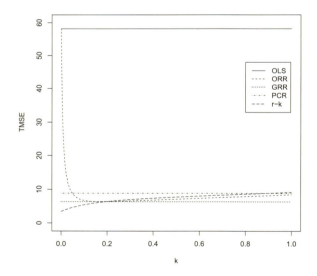

Figure 6.6:   TMSEs of OLS type $\widehat{\boldsymbol{\theta}}$, ORR type $\widehat{\boldsymbol{\theta}}(\cdot, k_0^*)$, GRR type $\widehat{\boldsymbol{\theta}}(k_0^*, \mathbf{K}_p^*)$, PCR type $\widehat{\boldsymbol{\theta}}(r^*, \cdot)$, and r-k class type $\widehat{\boldsymbol{\theta}}(r^*, k_0^*)$ estimators versus ridge coefficient k: Data Set 2

Chapter 7
Applied Feasible GRR Estimation to Linear Basis Function Models

In this chapter, we apply a feasible GRR (say FGRR) estimator to a linear basis function model. A linear basis function model (e.g. Bishop (2006)) has its mean structure modeled as a linear combination of suitable basis functions of some independent variables, and we are able to flexibly fit them to the data. The most popular estimating method of the weights in such models is the least squares method; however, some regularized versions of the least squares method are also applied to an over-fitting problem in the fields found in statistical learning and machine learning. Note that the exact moments and the MSE criteria of the FGRR estimator given in Chapter 5 are used.

7.1 Introduction

Suppose that the $(d+1)$-dimensional sample vector $[\boldsymbol{X}', Y]'$ has the joint distribution function $F(\boldsymbol{x}, y)$. If the sample Y is estimated by some function $\mathsf{r}(\boldsymbol{X})$ of the vector of sample $\boldsymbol{X}$, then the estimating problem is called the *statistical regression problem*:

$$Y \overset{\text{estimate}}{\longleftarrow} \mathsf{r}(\boldsymbol{X})$$

(e.g. Györfi *et al.* (2002)). The function $\mathsf{r}(\boldsymbol{X})$ is called the *regression function* and we assume that $\mathsf{r} \in L^p$ [1]. The sample Y is called a response variable, and $\boldsymbol{X} = [X_1, \ldots, X_d]'$ is a vector of explanatory variables. An observed vector of $\boldsymbol{X}$ is denoted by $\boldsymbol{x} = [x_1, \ldots, x_d]'$.

In the statistical regression problem, an important aspect is the selection method for the regression function r. A popular method is formulated by the following minimization problem:

$$\rho_2(Y, \mathsf{r}(\boldsymbol{X})) \longrightarrow \min_{\mathsf{r} \in L^2},$$

where ρ_2 is the mean squared error [2],

$$\rho_2(Y, \mathsf{r}(\boldsymbol{X})) := \int_{\mathbb{R}^{d+1}} |y - \mathsf{r}(\boldsymbol{x})|^2 dF(\boldsymbol{x}, y).$$

The solution to this minimization problem is the *conditional mean* of Y given by $\boldsymbol{X} = \boldsymbol{x} (\in \mathbb{R}^d)$:

$$\mathsf{r}^*(\boldsymbol{X}) = \mathrm{E}(Y \mid \boldsymbol{x}) =: \mu(\boldsymbol{x})$$

---
[1] L^p is the Lebesgue space (e.g. Stein and Shakarchi (2011)).
[2] It is also called L^2-risk.

(e.g. Györfi et al. (2002)). Therefore, we have to know the conditional mean $\mu(\boldsymbol{x})$ as a *mean structure* in the statistical regression problem. However, it is usually unknown because the joint distribution $F(\boldsymbol{x}, y)$ is not clear in general. As the first step, let us consider that the conditional mean $\mu(\boldsymbol{x})$ can be approximated by the following linear function of the observed vector of the explanatory variables $\boldsymbol{x} = [x_1, \ldots, x_d]'$:

$$\eta := \beta_0 + \boldsymbol{\beta}'\boldsymbol{x} = \beta_0 + \beta_1 x_1 + \cdots + \beta_d x_d \text{ (linear predictor)}$$

If we take
$$\mu(\boldsymbol{x}) \doteq \eta = \beta_0 + \beta_1 x_1 + \cdots + \beta_d x_d,$$
then the following linear regression model is obtained:

$$Y = \beta_0 + \beta_1 x_1 + \cdots + \beta_d x_d + \varepsilon, \quad (7.1)$$

where $\beta_0, \beta_1, \ldots, \beta_d$ are the regression coefficients, and the error term ε is represented by the difference between the response variable Y and the linear predictor η [3].

The linear regression model (7.1) is the most basic regression model and is easily treated. However, it is also a linear function of the explanatory variables x_i, and this imposes significant limitations on the model. (See Bishop (2006).) We treat *linear basis function* (LBF) models which are often used for this problem in the fields of *statistical learning* or *machine learning*. (e.g. Hastie et al. (2009), and Bishop (2006)) The LBF models are extensions of the linear regression model and they can be flexibly fit to a variety of data sets. Note that they include not only the linear regression model but also polynomial regression models, B-spline models, radian basis function models, and so on.

In such fields, there is a problem of estimation known as *over-fitting* which is similar to multicollinearity with respect to linear regression analysis. Some *regularized least squares methods* (e.g. Bishop (2006), and Konishi and Kitagawa (2007)) have been proposed for estimating parameters, for example, the ORR estimator by Hoerl and Kennard (1970a) and the *least absolute shrinkage and selection operator* (LASSO) estimator by Tibshirani (1996). However, the moments with respect to a feasible version of the estimators have not been discussed in these fields.

In this chapter, we apply FGRR estimation to the LBF models. The reason for adopting the FGRR estimator is that its exact moments are known as given in Chapter 5 [4].

7.2 Models and Estimators

---

[3] The linear predictor is called the *structural part* of the model (7.1), and the error term is referred to as the *stochastic part*.

[4] Note that the generalized ridge regression (GRR) estimator can be derived from a regularized least squares method. (See Section 6 (Remarks) in Inagaki (1977).)

7.2. MODELS AND ESTIMATORS

Suppose that the conditional expectation $\mu(\boldsymbol{x}) = \mathrm{E}(Y \mid \boldsymbol{x})$ of the response variable Y given a vector of explanatory variables $\boldsymbol{X} = \boldsymbol{x}(\in \mathbb{R}^d)$ can be approximated by the following linear combination of known functions $b_j(\boldsymbol{x})$ ($j = 0, 1, \ldots, p$) called *basis functions*:

$$\mu(\boldsymbol{x}) \simeq w_0 + w_1 b_1(\boldsymbol{x}) + \cdots + w_p b_p(\boldsymbol{x}) = \boldsymbol{w}' \boldsymbol{b}(\boldsymbol{x}),$$

where $\boldsymbol{w} := [w_0, w_1, \ldots, w_p]'$ is an unknown vector of parameters and is called a *weight vector*, and $\boldsymbol{b}(\boldsymbol{x}) := [b_0(\boldsymbol{x}), b_1(\boldsymbol{x}), \ldots, b_p(\boldsymbol{x})]' = [1, b_1(\boldsymbol{x}), \ldots, b_p(\boldsymbol{x})]'$ is called a *linear basis function vector* (known).

The following model is called the *linear basis function* (LBF) model:

$$Y = w_0 + w_1 b_1(\boldsymbol{x}) + \cdots + w_p b_p(\boldsymbol{x}) + \varepsilon.$$

See Bishop (2006). The LBF model can also be represented by taking the index i of the individual as follows:

$$Y_i = w_0 + w_1 b_1(\boldsymbol{x}_i) + \cdots + w_p b_p(\boldsymbol{x}_i) + \varepsilon_i, \quad i = 1, \ldots, n.$$

Furthermore, we can represent it in matrix-vector form as follows:

$$\boldsymbol{Y} = \boldsymbol{B}\boldsymbol{w} + \boldsymbol{\varepsilon}, \tag{7.2}$$

where

$$\boldsymbol{Y} := \begin{bmatrix} Y_1 \\ \vdots \\ Y_n \end{bmatrix}, \quad \boldsymbol{B} := \begin{bmatrix} b_0(\boldsymbol{x}_1) & b_1(\boldsymbol{x}_1) & \cdots & b_p(\boldsymbol{x}_1) \\ \vdots & \vdots & & \vdots \\ b_0(\boldsymbol{x}_n) & b_1(\boldsymbol{x}_n) & \cdots & b_p(\boldsymbol{x}_n) \end{bmatrix},$$

$$\boldsymbol{w} := \begin{bmatrix} w_0 \\ w_1 \\ \vdots \\ w_p \end{bmatrix}, \quad \boldsymbol{\varepsilon} := \begin{bmatrix} \varepsilon_1 \\ \vdots \\ \varepsilon_n \end{bmatrix}.$$

Furthermore, we set $b_0(\boldsymbol{x}_i) := 1$ and suppose that $\mathrm{rank}(\boldsymbol{B}) = p + 1 (< n)$, $\boldsymbol{\varepsilon} \sim N_n(\boldsymbol{0}, \sigma^2 \boldsymbol{I}_n)$.

Remark 7.1 (Variations of LBF models). *There are several variations of LBF models. For example, if the power function*

$$b_j(x) = x^j$$

is used as the linear basis function, then the LBF model is the following polynomial regression model:

$$Y_i = \sum_{j=0}^{p} w_j b_j(x_i) + \varepsilon_i = w_0 + \sum_{j=1}^{p} w_j x_i^j + \varepsilon_i.$$

7 APPLIED FEASIBLE GRR ESTIMATION TO LINEAR BASIS FUNCTION MODELS

Another possibility is the 'Gaussian' *basis function of the form*

$$b_j(\boldsymbol{x}) = \exp\left(-\frac{\|\boldsymbol{x} - \boldsymbol{\mu}_j\|^2}{2s_j^2}\right), \quad j = 1, \ldots, p,$$

where $\boldsymbol{\mu}_j$ is the location parameter vector of the basis functions, and the parameters s_j are their scales. Then, the linear basis function model is

$$Y_i = w_0 + \sum_{j=1}^{p} w_j b_j(\boldsymbol{x}_i) + \varepsilon_i = w_0 + \sum_{j=1}^{p} w_j \exp\left(-\frac{\|\boldsymbol{x}_i - \boldsymbol{\mu}_j\|^2}{2s_j^2}\right) + \varepsilon_i,$$

which is called the radial basis function model.

The OLS estimator of $\boldsymbol{w}$ is given by

$$\widehat{\boldsymbol{w}} := (\mathbf{B}'\mathbf{B})^{-1}\mathbf{B}'\boldsymbol{Y}. \tag{7.3}$$

It has been pointed out that the estimation (or predictive) accuracy of the OLS estimator is wrong, caused by the over-fitting problem. (See Bishop (2006) and Konishi and Kitagawa (2007).) Note that the problem is essentially equivalent to the multicollinearity problem caused by over-defined explanatory variables in the linear regression model.

Let us consider the estimating problem of the weight vector $\boldsymbol{w}$ by using the GRR estimator:

$$\widehat{\boldsymbol{w}}(\mathbf{K}) := (\mathbf{B}'\mathbf{B} + \boldsymbol{\Gamma}\mathbf{K}\boldsymbol{\Gamma}')^{-1}\mathbf{B}'\boldsymbol{Y}, \tag{7.4}$$

where $\boldsymbol{\Gamma}$ is an orthogonal matrix such that

$$\boldsymbol{\Gamma}'\mathbf{B}'\mathbf{B}\boldsymbol{\Gamma} = \boldsymbol{\Lambda} = \mathrm{diag}(\lambda_0, \lambda_1, \ldots, \lambda_p),$$

$\lambda_0 \geq \lambda_1 \geq \cdots \geq \lambda_p(>0)$ are the eigenvalues of the matrix $\mathbf{B}'\mathbf{B}$, and

$$\mathbf{K} := \mathrm{diag}(k_0, k_1, \ldots, k_p)$$

is a diagonal matrix with the ridge coefficients $k_i (\geq 0)$, $i = 0, 1, \ldots, p$.

Next, we treat the canonical form of the LBF model as follows:

$$\boldsymbol{Y} = \mathbf{A}\boldsymbol{\alpha} + \boldsymbol{\varepsilon}, \tag{7.5}$$

where

$$\mathbf{A} := \mathbf{B}\boldsymbol{\Gamma},$$

and the relationship

$$\boldsymbol{\alpha} := \boldsymbol{\Gamma}'\boldsymbol{w} \tag{7.6}$$

holds. Note that we treat the canonical form (7.5) which is given by the orthogonal transformation $\boldsymbol{\Gamma}$ of the matrix $\mathbf{B}$ without standardization.

The OLS estimator of $\boldsymbol{\alpha}$ in the canonical form (7.5) is

$$\widehat{\boldsymbol{\alpha}} := (\mathbf{A}'\mathbf{A})^{-1}\mathbf{A}'\mathbf{Y} = \boldsymbol{\Lambda}^{-1}\mathbf{A}'\mathbf{Y}, \tag{7.7}$$

and the GRR estimator is defined by

$$\widehat{\boldsymbol{\alpha}}(\mathbf{K}) := (\mathbf{A}'\mathbf{A} + \mathbf{K})^{-1}\mathbf{A}'\mathbf{Y} = (\boldsymbol{\Lambda} + \mathbf{K})^{-1}\mathbf{A}'\mathbf{Y}. \tag{7.8}$$

Remark 7.2. *The relationship (7.6) between the parameter vectors $\boldsymbol{w}$ and $\boldsymbol{\alpha}$*

$$\boldsymbol{\alpha} := \boldsymbol{\Gamma}'\boldsymbol{w} \iff \boldsymbol{w} = \boldsymbol{\Gamma}\boldsymbol{\alpha}$$

implies the following relationship between their estimators:

$$\widehat{\boldsymbol{w}} = \boldsymbol{\Gamma}\widehat{\boldsymbol{\alpha}}, \quad \widehat{\boldsymbol{w}}(\mathbf{K}) = \boldsymbol{\Gamma}\widehat{\boldsymbol{\alpha}}(\mathbf{K}).$$

Thus, the OLS estimator and the GRR estimator have invariances with respect to the orthogonal transformation $\boldsymbol{\Gamma}$. (See also Section 1.3.)

Form the results of Section 3.4, the optimum values of ridge coefficients k_i, $(i = 0, 1, \ldots, p)$ of the GRR estimator are given by

$$k_i^* = \frac{\sigma^2}{\alpha_i^2}, \quad i = 0, 1, \ldots, p,$$

and their estimators are

$$\widehat{k}_i^* = \frac{\widehat{\sigma}^2}{\widehat{\alpha}_i^2}, \quad i = 0, 1, \ldots, p.$$

See Section 4.2.

We can obtain the FGRR estimator for $\boldsymbol{w}$ by plugging $\widehat{k}_i^*$ in the GRR estimator $\widehat{\boldsymbol{w}}(\mathbf{K})$ as follows:

$$\widehat{\boldsymbol{w}}^* := \widehat{\boldsymbol{w}}(\widehat{\mathbf{K}}^*) = (\mathbf{B}'\mathbf{B} + \boldsymbol{\Gamma}\widehat{\mathbf{K}}^*\boldsymbol{\Gamma}')^{-1}\mathbf{B}'\mathbf{Y}, \tag{7.9}$$

where $\widehat{\mathbf{K}}^* := \mathrm{diag}(\widehat{k}_0^*, \widehat{k}_1^*, \ldots, \widehat{k}_p^*)$.

Similarly, the FGRR estimator for $\boldsymbol{\alpha}$ is given by

$$\widehat{\boldsymbol{\alpha}}^* := \widehat{\boldsymbol{\alpha}}(\widehat{\mathbf{K}}^*) = \left(\mathbf{A}'\mathbf{A} + \widehat{\mathbf{K}}^*\right)^{-1}\mathbf{A}'\mathbf{Y} = \left(\boldsymbol{\Lambda} + \widehat{\mathbf{K}}^*\right)^{-1}\mathbf{A}'\mathbf{Y}. \tag{7.10}$$

Remark 7.3. *Note that the FGRR estimator for $\boldsymbol{w}$ in the LBF model (7.2) is given by using the orthogonal transformation $\boldsymbol{\Gamma}$ of $\widehat{\boldsymbol{\alpha}}^*$ as follows:*

$$\widehat{\boldsymbol{w}}^* = \boldsymbol{\Gamma}\widehat{\boldsymbol{\alpha}}^*. \tag{7.11}$$

Therefore, the FGRR estimator also has invariance with respect to the orthogonal transformation. See also Remark 7.2.

7.3 Exact Moments and Mean Square Criteria of FGRR Estimator

From the results in Chapter 5, the mean vector and variance-covariance matrix of the FGRR estimator $\widehat{\boldsymbol{\alpha}}^*$ are given as follows:

$$\mathrm{E}(\widehat{\boldsymbol{\alpha}}^*) = \mathcal{A}\boldsymbol{\mu}_1, \quad \mathrm{V}(\widehat{\boldsymbol{\alpha}}^*) = \mathcal{A}(\mathbf{M}_2 - \boldsymbol{\mu}_1\boldsymbol{\mu}_1')\mathcal{A}, \tag{7.12}$$

where $\mathcal{A} := \mathrm{diag}(\alpha_0, \alpha_1, \ldots, \alpha_p)$, and

$$\boldsymbol{\mu}_1 := \begin{bmatrix} \mu_{10} \\ \mu_{11} \\ \vdots \\ \mu_{1p} \end{bmatrix}, \quad \mathbf{M}_2 := \begin{bmatrix} \mu_{20} & \mu_{1101} & \cdots & \mu_{110p} \\ \mu_{1110} & \mu_{21} & \cdots & \mu_{111p} \\ \vdots & \vdots & \ddots & \vdots \\ \mu_{11p0} & \mu_{11p1} & \cdots & \mu_{2p} \end{bmatrix},$$

$$\mu_{1i} := \mu_1(\delta_i, \nu) = \sum_{\ell=0}^{\infty} \left(\frac{\nu-1}{\nu}\right)^{\ell} \sum_{m=0}^{\infty} \frac{B\left(\ell + \frac{\nu}{2}, m + \frac{5}{2}\right)}{B\left(\frac{\nu}{2}, m + \frac{3}{2}\right)} p(m; \delta_i), \tag{7.13}$$

$$\mu_{2i} := \mu_2(\delta_i, \nu) = \frac{1}{2\delta_i}\sum_{\ell=0}^{\infty}(\ell+1)\left(\frac{\nu-1}{\nu}\right)^{\ell} \sum_{m=0}^{\infty} (2m+1)\frac{B\left(\ell + \frac{\nu}{2}, m + \frac{7}{2}\right)}{B\left(\frac{\nu}{2}, m + \frac{3}{2}\right)} p(m; \delta_i), \tag{7.14}$$

$$\mu_{11ij} := \mu_{11}(\delta_i, \delta_j, \nu) = \sum_{\ell_1=0}^{\infty}\sum_{\ell_2=0}^{\infty}\left(\frac{\nu-1}{\nu}\right)^{\ell_1+\ell_2}$$
$$\times \sum_{m_1=0}^{\infty}\sum_{m_2=0}^{\infty} \frac{B_{\ell_1+1\,\ell_2+1}\left(m_1+\frac{5}{2}, m_2+\frac{5}{2}, \ell_1+\ell_2+\frac{\nu}{2}\right)}{B\left(m_1+\frac{3}{2}, m_2+\frac{3}{2}, \frac{\nu}{2}\right)} p(m_1;\delta_i)p(m_2;\delta_j). \tag{7.15}$$

Since the FGRR estimator has invariance (7.11), the mean vector and variance-covariance matrix of the FGRR estimator $\widehat{\boldsymbol{w}}^*$ are given by

$$\mathrm{E}(\widehat{\boldsymbol{w}}^*) = \boldsymbol{\Gamma}\mathcal{A}\boldsymbol{\mu}_1, \quad \mathrm{V}(\widehat{\boldsymbol{w}}^*) = \boldsymbol{\Gamma}\mathcal{A}(\mathbf{M}_2 - \boldsymbol{\mu}_1\boldsymbol{\mu}_1')\mathcal{A}\boldsymbol{\Gamma}',$$

respectively.

Next, we consider the mean square error (MSE) criteria of the FGRR estimator. From (7.12), the MSE matrix of $\widehat{\boldsymbol{\alpha}}^*$ is given by

$$\begin{aligned}\mathrm{MSE}(\widehat{\boldsymbol{\alpha}}^*) &= \mathrm{V}(\widehat{\boldsymbol{\alpha}}^*) + \mathrm{bias}(\widehat{\boldsymbol{\alpha}}^*)\mathrm{bias}(\widehat{\boldsymbol{\alpha}}^*)' \\ &= \mathcal{A}(\mathbf{M}_2 - \boldsymbol{\mu}_1\boldsymbol{\mu}_1')\mathcal{A} + (\mathbf{M}_1 - \mathbf{I}_{p+1})\boldsymbol{\alpha}\boldsymbol{\alpha}'(\mathbf{M}_1 - \mathbf{I}_{p+1}),\end{aligned} \tag{7.16}$$

where $\mathbf{M}_1 := \mathrm{diag}(\mu_{10}, \mu_{11}, \ldots, \mu_{1p})$, and the (i,j) components of the MSE matrix are

$$\left[\mathrm{MSE}\left(\widehat{\boldsymbol{\alpha}}^*\right)\right]_{ij} = \begin{cases} \alpha_i^2\left(\mu_{2i} - 2\mu_{1i} + 1\right) = \mathrm{MSE}\left(\widehat{\alpha}_i^*\right) & ; \text{if } i = j, \\ \alpha_i\alpha_j\left\{\mu_{11ij} - (\mu_{1i} + \mu_{1j}) + 1\right\} = \mathrm{MCE}\left(\widehat{\alpha}_i^*, \widehat{\alpha}_j^*\right) & ; \text{if } i \neq j. \end{cases} \tag{7.17}$$

The total mean square error (TMSE) of $\widehat{\boldsymbol{\alpha}}^*$ is given by the trace of the MSE matrix MSE $(\widehat{\boldsymbol{\alpha}}^*)$ as follows:

$$\text{TMSE}\left(\widehat{\boldsymbol{\alpha}}^*\right) := \sum_{i=0}^{p} \text{MSE}(\widehat{\alpha}_i(\widehat{k}_i^*)) = trace\,\text{MSE}\left(\widehat{\boldsymbol{\alpha}}^*\right) = \sum_{i=0}^{p} \alpha_i^2 \left(\mu_{2i} - 2\mu_{1i} + 1\right). \tag{7.18}$$

Remark 7.4. *Since the FGRR estimator has invariance (7.11),*

$$\begin{aligned}
TMSE(\widehat{\boldsymbol{w}}^*) &= trace\ V(\widehat{\boldsymbol{w}}^*) + \|bias(\widehat{\boldsymbol{w}}^*)\|^2. \\
&= trace\ V(\boldsymbol{\Gamma}\widehat{\boldsymbol{\alpha}}^*) + \|bias(\boldsymbol{\Gamma}\widehat{\boldsymbol{\alpha}}^*)\|^2. \\
&= trace\ \boldsymbol{\Gamma}\,V(\widehat{\boldsymbol{\alpha}}^*)\boldsymbol{\Gamma}' + \|\boldsymbol{\Gamma}\,bias(\widehat{\boldsymbol{\alpha}}^*)\|^2. \\
&= TMSE(\widehat{\boldsymbol{\alpha}}^*).
\end{aligned}$$

That is, we can use the canonical form (7.5) instead of the LBF model (7.2) only if we evaluate the FGRR estimator by the TMSE.

7.4 Numerical Example: Polynomial Regression Model

Let us give an example of the LBF model [5]. Suppose that the conditional expectation of the response variable Y given $X = x$ is

$$\mathrm{E}(Y \mid x) = \sin(x) : \text{ sine function.}$$

Since the sine function is approximated by the polynomial function

$$\begin{aligned}
\sin(x) &\simeq x - \frac{1}{6}x^3 + \frac{1}{120}x^5 - \frac{1}{5040}x^7 + \frac{1}{362880}x^9 \\
&= x - 0.166667x^3 + 0.008333x^5 - 0.000198x^7 + 0.000003x^9,
\end{aligned}$$

which is the *Taylor series expansion* of order 9 at $x = 0$, we consider the following *polynomial regression model* as the LBF model:

$$\begin{aligned}
y &= \sum_{j=0}^{p} w_j b_j(x) + \epsilon = \sum_{j=0}^{p} w_j x^j + \epsilon \\
&= x - 0.166667x^3 + 0.008333x^5 - 0.000198x^7 + 0.000003x^9 + \epsilon, \tag{7.19}
\end{aligned}$$

where the linear basis function is $b_j(x) = x^j$ (power function) and $p = 9$. (See Remark 7.1.)

Let us consider the data set (Table 7.1) generated by the model (7.19) [6].

---
[5] We refer the reader to Chapter 1 of Bishop (2006) for more details.
[6] The error term is i.i.d. $N(0, \sigma^2)$, where $\sigma^2 = 0.5^2$.

Table 7.1: Data Set: Polynomial Regression Model

i	1	2	3	4	5	6	7	8	9	10	11
x_i	-3.14	-2.51	-1.88	-1.26	-0.63	0.00	0.63	1.26	1.88	2.51	3.14
y_i	-0.68	-0.64	-1.46	-1.09	-1.29	0.62	0.52	1.22	1.00	0.66	-0.45

First of all, if we fit the full model (the polynomial regression model of order 10)

$$y_i = \sum_{j=0}^{10} w_j x^j + \epsilon_i,$$

then the sample regression curve estimated by the OLS estimators $\widehat{w}_j$ is derived from the data set (Table 7.1) as follows:

$$\begin{aligned} y &= \widehat{w}_0 + \widehat{w}_1 x + \widehat{w}_2 x^2 + \cdots + \widehat{w}_{10} x^{10} \\ &= 0.624 + 1.736x - 3.996x^2 - 0.847x^3 + 4.229x^4 \\ &\quad + 0.266x^5 - 1.535x^6 - 0.039x^7 + 0.217x^8 + 0.002x^9 - 0.010x^{10}. \end{aligned} \quad (7.20)$$

The fitted curve (7.20) passes through all data points. (See Figure 7.1.) However, it fluctuates widely and cannot explain the mean structure $\sin(x)$. This result is a typical example of over-fitting.

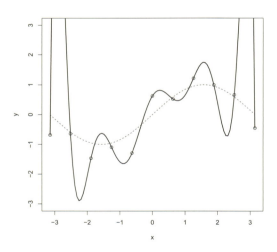

Figure 7.1: Sample regression curve (solid line) of order 10 estimated by the OLS estimators $\widehat{w}_j$ and sine curve (dash line)

7.4. NUMERICAL EXAMPLE: POLYNOMIAL REGRESSION MODEL

Next, we fit the polynomial regression model of order 9 (true order) by the method of least squares. The sample regression curve is obtained as follows:

$$\begin{aligned} y &= \widehat{w}_0 + \widehat{w}_1 x + \widehat{w}_2 x^2 + \cdots + \widehat{w}_9 x^9 \\ &= 0.138 + 1.736x - 0.338x^2 - 0.847x^3 + 0.108x^4 \\ &\quad + 0.266x^5 - 0.011x^6 - 0.039x^7 + 0.000x^8 + 0.002x^9. \end{aligned} \quad (7.21)$$

Figure 7.2 is a scatter plot with the sample regression curve (7.21). Note that it is more stable than the case of order 10; however, it fluctuates slightly.

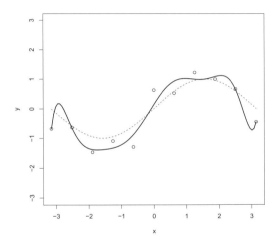

Figure 7.2: Sample regression curve (solid line) of order 9 estimated by the OLS estimators $\widehat{w}_j$ and sine curve (dash line)

Finally, we give the sample regression curve of order 9 estimated by the FGRR estimators $\widehat{w}_j^*$ as follows:

$$\begin{aligned} y &= \widehat{w}_0^* + \widehat{w}_1^* x + \widehat{w}_2^* x^2 + \cdots + \widehat{w}_9^* x^9 \\ &= 0.004 + 0.869x - 0.016x^2 + 0.041x^3 + 0.002x^4 \\ &\quad - 0.061x^5 + 0.000x^6 + 0.008x^7 + 0.000x^8 + 0.000x^9. \end{aligned} \quad (7.22)$$

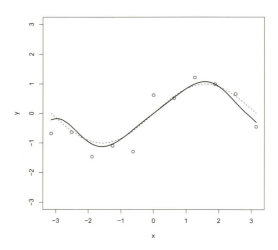

Figure 7.3: Sample regression curve (solid line) of order 9 estimated by the FGRR estimators $\widehat{w}_j^*$ and sine curve (dash line)

The scatter plot with the regression curve (7.22) is given by Figure 7.3. The curve (7.22) seems very good in terms of being stable and being able to explain the mean structure $\sin(x)$. The reason which the FGRR estimation works well is explained as follows:

- From Remark 5.7 and Table 5.4,

$$\delta_i = \frac{1}{2}\frac{\alpha_i^2}{\sigma^2/\lambda_i} < 1 \Longrightarrow \mathrm{MSE}(\widehat{\alpha}_i^*) < \mathrm{MSE}(\widehat{\alpha}_i). \qquad (7.23)$$

- $\sigma^2 = 0.5^2 = 0.25$, $\nu = n - p - 1 = 11 - 9 - 1 = 1$. See the MSEs of the OLS estimators and the FGRR estimators and some parameters listed in Table 7.2; note that some δ_i are less than one, and that in these cases the MSEs of the FGRR estimators are less than those of the OLS estimator from (7.23). Additionally, the TMSE (total sum of the MSEs) of the FGRR estimators is less than that of the OLS estimators:

$$\mathrm{TMSE}(\widehat{\boldsymbol{\alpha}}^*) = 2.994756 < 4.495567 = \mathrm{TMSE}(\widehat{\boldsymbol{\alpha}}).$$

Recall that $\mathrm{TMSE}(\widehat{\boldsymbol{w}}) = \mathrm{TMSE}(\widehat{\boldsymbol{\alpha}})$ and $\mathrm{TMSE}(\widehat{\boldsymbol{w}}^*) = \mathrm{TMSE}(\widehat{\boldsymbol{\alpha}}^*)$ hold, and consequently

$$\mathrm{TMSE}(\widehat{\boldsymbol{w}}^*) = 2.994756 < 4.495567 = \mathrm{TMSE}(\widehat{\boldsymbol{w}}).$$

Table 7.2: MSEs of OLS and FGRR estimators and some parameters

i	$1/\lambda_i$	α_i	δ_i	MSE($\widehat{\alpha}_i$)	MSE($\widehat{\alpha}_i^*$)
0	0.000000	-0.000003	0.036501	0.000000	0.000000
1	0.000000	0.000000	0.000000	0.000000	0.000000
2	0.000009	0.003389	2.637165	0.000002	0.000003
3	0.000053	0.000000	0.000000	0.000013	0.000008
4	0.008363	0.140151	4.697371	0.002091	0.002580
5	0.026309	0.000000	0.000000	0.006577	0.004111
6	0.304886	0.000000	0.000000	0.076221	0.047638
7	0.443929	0.754144	2.562271	0.110982	0.133129
8	4.633144	0.000000	0.000000	1.158286	0.723929
9	12.565575	0.662918	0.069947	3.141394	2.083358

Chapter 8
Concluding Remarks

It has been pointed out that if there exists multicollinearity among the vectors of the explanatory variables in a regression model, then several problems occur. These problems concern the accuracy of the ordinary least squares estimator for estimating the regression coefficients. In this book, we improved the accuracy of estimation by using several alternative shrinkage regression estimators, that is, the ordinary ridge regression estimator, the generalized ridge regression estimator, the principal component regression estimator, and the r-k class estimator. We adopted the total mean squared error as a measure of the improvement. We also considered the feasibilities of these estimators.

There are the following problems with respect to using the shrinkage regression estimators:

(P-SRE 1) The shrinkage regression estimators have customarily been treated under the standardized model (M_X). However, it is important in applications that we consider the shrinkage regression estimators under the original (usual) model (M_Z).

(P-SRE 2) Although many algorithms for choosing the number of principal components and the ridge coefficients have been proposed, these depend on unknown parameters, which need to be estimated. The concrete algorithms replace the unknown parameters with these estimators. To use the feasible shrinkage regression estimators, we need to plug them into the appropriate parts; however, it is difficult to obtain the total mean squared error analytically.

For problem (P-SRE 1), we considered several shrinkage-type regression estimators under the usual model (M_Z) in Chapter 2. We also compared the total mean squared errors of the estimators in Chapter 3. For problem (P-SRE 2), we gave a variety of algorithms for choosing the number of principal components and the ridge coefficients and considered the feasible shrinkage regression estimators in Chapter 4. We also precisely investigated the exact moments and mean squared error criteria of the feasible generalized ridge regression estimator in Chapter 5. We can use the results as a "benchmark", and then it is possible to know the improvement of the shrinkage(-type) regression estimators by comparing the results of several Monte Carlo studies by means of a parametric bootstrap method under multicollinearity. (See Chapter 6.) These results are important for applications because we can objectively evaluate the total mean squared error of the feasible shrinkage regression estimators from only a data set.

In recent years, some shrinkage regression estimators have been used in the fields found in statistical learning and machine learning. In Chapter 7, we considered the linear basis function model, which is basic in these fields, and estimated the coefficients (weights) of the model by using the feasible generalized ridge regression estimators in the over-fitting problem. The reason for adopting the feasible generalized ridge regression estimator was that its exact moments were known, as given in Chapter 5. A numerical example based on a polynomial regression model was also given, and we could verify the degree of advantage of the feasible generalized ridge regression estimators over the ordinary least squares estimator.

In conclusion, if the (feasible) shrinkage regression estimators are appropriately used under multicollinearity, they are potentially more *efficient* than the ordinary least squares estimator. Finally, we hope that this book will be helpful in investigating the shrinkage regression estimators and their feasibilities.

Appendix A
Special Functions

A.1 Gamma Function

Definition A.1 (Gamma function). The *gamma function* is defined by
$$\Gamma(x) := \int_0^\infty e^{-t} t^{x-1} dt, \quad x > 0.$$

Definition A.2 (Digamma function). The *digamma function* is defined by
$$\psi(x) := (\log \Gamma(x))' = \frac{\Gamma'(x)}{\Gamma(x)}.$$

Proposition A.1. *The gamma function has the following properties:*

(a) $\Gamma(x+1) = x\Gamma(x)$

(b) $\Gamma(n+1) = n!, \, \forall n \in \mathbb{N}$

(c) $\Gamma\left(\frac{1}{2}\right) = \sqrt{\pi}$

(d) $\Gamma^{(n)}(x) := \dfrac{d^n}{dx^n}\Gamma(x) = \int_0^\infty t^{x-1} e^{-t} (\log t)^n \, dt$

Proposition A.2. *The digamma function $\psi(x)$ is monotonically increasing. That is, the function $\log \Gamma(x)$ is convex.*

Proof. Note that
$$\Gamma(x)u^2 + 2\Gamma'(x)u + \Gamma''(x) = \int_0^\infty e^{-t} t^{x-1} u^2 dt + \int_0^\infty 2e^{-t} t^{x-1} u \log t \, dt$$
$$+ \int_0^\infty e^{-t} t^{x-1} (\log t)^2 \, dt$$
$$= \int_0^\infty e^{-t} t^{x-1} \left\{ u^2 + 2\log t + (\log t)^2 \right\} dt$$
$$= \int_0^\infty e^{-t} t^{x-1} (u + \log t)^2 \, dt \geq 0, \quad \forall u \in \mathbb{R},$$

so the discriminant is
$$\frac{D}{4} = \Gamma'(x)^2 - \Gamma(x)\Gamma''(x) \leq 0.$$

Therefore, the derivative of the digamma function is
$$\psi'(x) = (\log \Gamma(x))'' = \left(\frac{\Gamma'(x)}{\Gamma(x)}\right)' = \frac{\Gamma''(x)\Gamma(x) - \Gamma'(x)^2}{\Gamma(x)^2} \geq 0.$$

Hence, the digamma function $\psi(x)$ is monotonically increasing. □

A.2 Beta Function

Definition A.3 (Beta function). The *beta function* is defined by
$$B(x,y) := \int_0^1 t^{x-1}(1-t)^{y-1} dt, \quad x, y > 0.$$

Proposition A.3. *The beta function has the following properties:*

(a) $B(x,y) = \dfrac{\Gamma(x)\Gamma(y)}{\Gamma(x+y)}$

(b) $B(x+1, y) = \dfrac{x}{x+y} B(x,y), \ B(x, y+1) = \dfrac{y}{x+y} B(x, y)$

(c) $B(x+1, y+1) = \dfrac{xy}{(x+y+1)(x+y)} B(x, y)$

Proposition A.4. *The beta function $B(x, y)$ is (partially) monotonically decreasing.*

Proof. We know that the partial derivative of the beta function $B(x, y)$ with respect to x is non-positive because of the monotonically increasing property of the digamma function $\psi(x)$:

$$\begin{aligned}\frac{\partial B(x,y)}{\partial x} &= \left(\frac{\Gamma(x)\Gamma(y)}{\Gamma(x+y)}\right)' = \frac{\Gamma'(x)\Gamma(y)\Gamma(x+y) - \Gamma(x)\Gamma(y)\Gamma'(x+y)}{\Gamma(x+y)^2} \\ &= \frac{\Gamma'(x)}{\Gamma(x)} B(x,y) - \frac{\Gamma'(x+y)}{\Gamma(x+y)} B(x,y) \\ &= (\psi(x) - \psi(x+y)) B(x,y) \le 0.\end{aligned}$$

Similarly, $\dfrac{\partial B(x,y)}{\partial y} \le 0$ also holds. $\square$

Definition A.4 (Generalized beta function). A *generalized beta function* can be defined by
$$B(x, y, z) := \iint_{\mathcal{D}} t_1^{x-1} t_2^{y-1} (1 - t_1 - t_2)^{z-1} dt_1 dt_2, \quad x, y, z > 0,$$

where the domain $\mathcal{D} := \{(t_1, t_2) \mid t_1, t_2 > 0, t_1 + t_2 < 1\}$ is a simplex.

Proposition A.5. *The generalized beta function has the following properties:*

(a) $B(x, y, z) = \dfrac{\Gamma(x)\Gamma(y)\Gamma(z)}{\Gamma(x + y + z)}$

(b) $B(x, y, z) = B(x, y) B(x + y, z)$

Proposition A.6. *The generalized beta function $B(x, y, z)$ is (partially) monotonically decreasing.*

Proof. The proof is similar to the case of the beta function. $\square$

Appendix B
Lemmas for Moments

Lemma B.1 (Dwivedi *et al.*). *If $Z \sim N(\tau, 1)$ and $V \sim \chi_\nu^2$ are mutually independent random variables, then the following are true:*
for an odd number p,
$$E\left(\frac{Z^p V^q}{(Z^2+V)^r}\right) = 2^{q-r+\frac{p-1}{2}} \tau \sum_{m=0}^{\infty} \frac{\Gamma\left(m+q-r+\frac{p+\nu+2}{2}\right)}{\Gamma\left(\frac{\nu}{2}\right)\Gamma\left(m+\frac{3}{2}\right)} B\left(q+\frac{\nu}{2}, m+\frac{p+2}{2}\right) p(m; \tau^2/2) \tag{B.1}$$

for an even number p,
$$E\left(\frac{Z^p V^q}{(Z^2+V)^r}\right) = 2^{q-r+\frac{p}{2}} \sum_{m=0}^{\infty} \frac{\Gamma\left(m+q-r+\frac{p+\nu+1}{2}\right)}{\Gamma\left(\frac{\nu}{2}\right)\Gamma\left(m+\frac{1}{2}\right)} B\left(q+\frac{\nu}{2}, m+\frac{p+1}{2}\right) p(m; \tau^2/2) \tag{B.2}$$

See Appendix in Dwivedi *et al.* (1980) for the proof.

Lemma B.2. *If $Z_1 \sim N(\tau_1, 1)$, $Z_2 \sim N(\tau_2, 1)$, and $V \sim \chi_\nu^2$ are mutually independent random variables and p_1, p_2 are odd integers, then*

$$\begin{aligned}
& E\left(\frac{Z_1^{p_1} V^{q_1}}{(Z_1^2+V)^{r_1}} \frac{Z_2^{p_2} V^{q_2}}{(Z_2^2+V)^{r_2}}\right) \\
& = \tau_1 \tau_2 2^{q_1+q_2-r_1-r_2+\frac{p_1+p_2}{2}-1} \\
& \quad \times \sum_{m_1=0}^{\infty} \sum_{m_2=0}^{\infty} \frac{\Gamma\left(m_1+m_2+q_1+q_2-r_1-r_2+\frac{p_1+p_2+\nu}{2}+2\right)}{\Gamma\left(m_1+\frac{3}{2}\right)\Gamma\left(m_2+\frac{3}{2}\right)\Gamma\left(\frac{\nu}{2}\right)} \\
& \quad \times \mathcal{B}_{r_1 r_2}\left(m_1+\frac{p_1+2}{2}, m_2+\frac{p_2+2}{2}, q_1+q_2+\frac{\nu}{2}\right) p(m_1; \tau_1^2/2) p(m_2; \tau_2^2/2),
\end{aligned} \tag{B.3}$$

where
$$\begin{aligned}
& \mathcal{B}_{r_1 r_2}\left(m_1+\frac{p_1+2}{2}, m_2+\frac{p_2+2}{2}, q_1+q_2+\frac{\nu}{2}\right) \\
& := \iint_{\mathcal{D}} \frac{1}{(1-t_1)^{r_2}} \frac{1}{(1-t_2)^{r_1}} t_1^{m_1+\frac{p_1+2}{2}-1} t_2^{m_2+\frac{p_2+2}{2}-1} (1-t_1-t_2)^{q_1+q_2+\frac{\nu}{2}-1} dt_1 dt_2.
\end{aligned}$$

Proof. The joint density function of random vector (Z_1, Z_2, V) is
$$\begin{aligned}
f(z_1, z_2, v) & := f_1(z_1) f_2(z_2) f_3(v) \\
& := \frac{1}{\sqrt{2\pi}} e^{-\frac{(z_1-\tau_1)^2}{2}} \cdot \frac{1}{\sqrt{2\pi}} e^{-\frac{(z_2-\tau_2)^2}{2}} \cdot \frac{1}{\Gamma\left(\frac{\nu}{2}\right) 2^{\frac{\nu}{2}}} v^{\frac{\nu}{2}-1} e^{-\frac{v}{2}},
\end{aligned} \tag{B.4}$$

where $f_1(z_1)$, $f_2(z_2)$, and $f_3(v)$ are the marginal density functions of Z_1, Z_2, and V, respectively, and

$$\Gamma(x) := \int_0^\infty t^{x-1} e^{-t} dt$$

is the gamma function. Note that $\Gamma\left(\frac{1}{2}\right) = \sqrt{\pi}$. The joint density function in (B.4) can also be expressed as

$$f(z_1, z_2, v) = \frac{1}{\Gamma\left(\frac{1}{2}\right)^2 \Gamma\left(\frac{\nu}{2}\right) 2^{\frac{\nu+2}{2}}} v^{\frac{\nu}{2}-1} \exp\left\{-\frac{(z_1-\tau_1)^2 + (z_2-\tau_2)^2 + v}{2}\right\}. \tag{B.5}$$

In (B.5),

$$\exp\left\{-\frac{(z_1-\tau_1)^2 + (z_2-\tau_2)^2 + v}{2}\right\}$$
$$= \exp\left(-\frac{z_1^2 - 2\tau_1 z_1 + \tau_1^2 + z_2^2 - 2\tau_2 z_1 + \tau_2^2 + v}{2}\right)$$
$$= \exp\left(-\frac{\tau_1^2 + \tau_2^2}{2}\right) \exp\left(-\frac{z_1^2 + z_2^2 + v}{2}\right) \exp(\tau_1 z_1) \exp(\tau_2 z_2),$$

and if we use the Taylor expansions

$$e^{\tau_1 z_1} = \sum_{m_1=0}^\infty \frac{(\tau_1 z_1)^{m_1}}{m_1!}, \quad e^{\tau_2 z_2} = \sum_{m_2=0}^\infty \frac{(\tau_2 z_2)^{m_2}}{m_2!},$$

then we have

$$f(z_1, z_2, v) = \frac{1}{\Gamma\left(\frac{1}{2}\right)^2 \Gamma\left(\frac{\nu}{2}\right) 2^{\frac{\nu+2}{2}}} v^{\frac{\nu}{2}-1} \exp\left(-\frac{\tau_1^2 + \tau_2^2}{2}\right) \exp\left(-\frac{z_1^2 + z_2^2 + v}{2}\right)$$
$$\times \exp(\tau_1 z_1) \exp(\tau_2 z_2)$$
$$= \frac{1}{\Gamma\left(\frac{1}{2}\right)^2 \Gamma\left(\frac{\nu}{2}\right) 2^{\frac{\nu+2}{2}}} v^{\frac{\nu}{2}-1} \exp\left(-\frac{\tau_1^2 + \tau_2^2}{2}\right) \exp\left(-\frac{z_1^2 + z_2^2 + v}{2}\right)$$
$$\times \sum_{m_1=0}^\infty \frac{(\tau_1 z_1)^{m_1}}{m_1!} \sum_{m_2=0}^\infty \frac{(\tau_2 z_2)^{m_2}}{m_2!}.$$

Therefore,

$$\mathrm{E}\left(\frac{Z_1^{p_1} V^{q_1}}{(Z_1^2+V)^{r_1}} \frac{Z_2^{p_2} V^{q_2}}{(Z_2^2+V)^{r_2}}\right)$$
$$= \frac{e^{-\frac{\tau_1^2+\tau_2^2}{2}}}{\Gamma\left(\frac{1}{2}\right)^2 \Gamma\left(\frac{\nu}{2}\right) 2^{\frac{\nu+2}{2}}} \sum_{m_1=0}^\infty \sum_{m_2=0}^\infty \frac{\tau_1^{m_1} \tau_2^{m_2}}{m_1! \, m_2!}$$
$$\times \int_0^\infty \int_{-\infty}^\infty \int_{-\infty}^\infty \frac{z_1^{p_1+m_1}}{(z_1^2+v)^{r_1}} \frac{z_2^{p_2+m_2}}{(z_2^2+v)^{r_2}} v^{q_1+q_2+\frac{\nu}{2}-1} e^{-\frac{z_1^2+z_2^2+v}{2}} dz_1 dz_2 dv. \tag{B.6}$$

Let p_1, p_2 be odd integers and write $p_1 = 2s_1 + 1$, $p_2 = 2s_2 + 1$, where s_1, s_2 are integers. Substituting these into (B.6),

$$E\left(\frac{Z_1^{p_1} V^{q_1}}{(Z_1^2+V)^{r_1}} \frac{Z_2^{p_2} V^{q_2}}{(Z_2^2+V)^{r_2}}\right)$$
$$= \frac{e^{-\frac{\tau_1^2+\tau_2^2}{2}}}{\Gamma\left(\frac{1}{2}\right)^2 \Gamma\left(\frac{\nu}{2}\right) 2^{\frac{\nu+2}{2}}} \sum_{m_1=0}^{\infty} \sum_{m_2=0}^{\infty} \frac{\tau_1^{m_1}}{m_1!} \frac{\tau_2^{m_2}}{m_2!}$$
$$\times \int_0^{\infty} v^{q_1+q_2+\frac{\nu}{2}-1} e^{-\frac{v}{2}} \int_{-\infty}^{\infty} g_1(z_1) dz_1 \int_{-\infty}^{\infty} g_2(z_2) dz_2 \, dv, \qquad (B.7)$$

where

$$g_j(z_j) := \frac{z_j^{2s_1+1+m_1}}{(z_j^2+v)^{r_j}} e^{-\frac{z_j^2}{2}}, \; j=1,2,$$

satisfies the following properties:

$$g_j(-z_j) = \begin{cases} -g_j(z_j) & ; \text{ if } m_j \text{ is even,} \\ g_j(z_j) & ; \text{ if } m_j \text{ is odd.} \end{cases}$$

If m_j is an even integer, then

$$\int_{-\infty}^{\infty} g_j(z_j) dz_j = 0.$$

If m_j is an odd integer, then

$$E\left(\frac{Z_1^{p_1} V^{q_1}}{(Z_1^2+V)^{r_1}} \frac{Z_2^{p_2} V^{q_2}}{(Z_2^2+V)^{r_2}}\right)$$
$$= \frac{e^{-\frac{\tau_1^2+\tau_2^2}{2}}}{\Gamma\left(\frac{1}{2}\right)^2 \Gamma\left(\frac{\nu}{2}\right) 2^{\frac{\nu+2}{2}}} \sum_{m_1=0}^{\infty} \sum_{m_2=0}^{\infty} \frac{\tau_1^{(2m_1+1)}}{(2m_1+1)!} \frac{\tau_2^{(2m_2+1)}}{(2m_2+1)!}$$
$$\times \int_0^{\infty} \int_{-\infty}^{\infty} \int_{-\infty}^{\infty} \frac{(z_1^2)^{s_1+m_1+1}}{(z_1^2+v)^{r_1}} \frac{(z_2^2)^{s_2+m_2+1}}{(z_2^2+v)^{r_2}} v^{q_1+q_2+\frac{\nu}{2}-1} e^{-\frac{z_1^2+z_2^2+v}{2}} dz_1 dz_2 dv.$$

From the duplication formula

$$\Gamma\left(\frac{1}{2}\right)^2 (2m_1+1)!(2m_2+1)! = 2^{2m_1+1} m_1! \Gamma\left(m_1+\frac{3}{2}\right) 2^{2m_2+1} m_2! \Gamma\left(m_2+\frac{3}{2}\right),$$

we have
$$\mathrm{E}\left(\frac{Z_1^{p_1}V^{q_1}}{(Z_1^2+V)^{r_1}}\frac{Z_2^{p_2}V^{q_2}}{(Z_2^2+V)^{r_2}}\right)$$
$$=\frac{e^{-\frac{\tau_1^2+\tau_2^2}{2}}\tau_1\tau_2}{\Gamma\left(\frac{\nu}{2}\right)2^{\frac{\nu+2}{2}}}\sum_{m_1=0}^{\infty}\sum_{m_2=0}^{\infty}\frac{(\tau_1^2)^{m_1}}{m_1!2^{2m_1+1}\Gamma\left(m_1+\frac{3}{2}\right)}\frac{(\tau_2^2)^{m_2}}{m_2!2^{2m_2+1}\Gamma\left(m_2+\frac{3}{2}\right)}$$
$$\times\int_0^{\infty}\int_{-\infty}^{\infty}\int_{-\infty}^{\infty}\frac{(z_1^2)^{s_1+m_1+1}}{(z_1^2+v)^{r_1}}\frac{(z_2^2)^{s_2+m_2+1}}{(z_2^2+v)^{r_2}}v^{q_1+q_2+\frac{\nu}{2}-1}e^{-\frac{z_1^2+z_2^2+v}{2}}dz_1dz_2dv$$
$$=\frac{\tau_1\tau_2}{\Gamma\left(\frac{\nu}{2}\right)2^{\frac{\nu+6}{2}}}\sum_{m_1=0}^{\infty}\sum_{m_2=0}^{\infty}\frac{p(m_1;\tau_1^2/2)}{2^{m_1}\Gamma\left(m_1+\frac{3}{2}\right)}\frac{p(m_2;\tau_2^2/2)}{2^{m_2}\Gamma\left(m_2+\frac{3}{2}\right)}R_{m_1m_2},\quad\text{(B.8)}$$

where $p(m_1;\tau_1^2/2)$, $p(m_2;\tau_2^2/2)$ are the probability mass functions of the Poisson distribution with parameters $\tau_1^2/2$, $\tau_2^2/2$, respectively, and

$$R_{m_1m_2}:=\int_0^{\infty}\int_{-\infty}^{\infty}\int_{-\infty}^{\infty}\frac{(z_1^2)^{s_1+m_1+1}}{(z_1^2+v)^{r_1}}\frac{(z_2^2)^{s_2+m_1+1}}{(z_2^2+v)^{r_2}}v^{q_1+q_2+\frac{\nu}{2}-1}e^{-\frac{z_1^2+z_2^2+v}{2}}dz_1dz_2dv$$
$$=\int_0^{\infty}v^{q_1+q_2+\frac{\nu}{2}-1}e^{-\frac{v}{2}}$$
$$\times\left\{\int_{-\infty}^{\infty}\frac{(z_1^2)^{s_1+m_1+1}}{(z_1^2+v)^{r_1}}e^{-\frac{z_1^2}{2}}dz_1\right\}\left\{\int_{-\infty}^{\infty}\frac{(z_2^2)^{s_2+m_2+1}}{(z_2^2+v)^{r_2}}e^{-\frac{z_2^2}{2}}dz_2\right\}dv.$$

Using the change of variables
$$u_1:=z_1^2\iff z_1=\begin{cases}-\sqrt{u_1},&\text{if }-\infty<z_1\le 0,\\\sqrt{u_1},&\text{if }0<z_1<\infty,\end{cases}$$
$$\implies dz_1=\begin{cases}-\frac{1}{2\sqrt{u_1}}du_1,&\text{if }-\infty<z_1\le 0,\\\frac{1}{2\sqrt{u_1}}du_1,&\text{if }0<z_1<\infty,\end{cases}$$

we obtain the following expression:
$$\int_{-\infty}^{\infty}\frac{(z_1^2)^{s_1+m_1+1}}{(z_1^2+v)^{r_1}}e^{-\frac{z_1^2}{2}}dz_1$$
$$=\int_{-\infty}^{0}\frac{(z_1^2)^{s_1+m_1+1}}{(z_1^2+v)^{r_1}}e^{-\frac{z_1^2}{2}}dz_1+\int_{0}^{\infty}\frac{(z_1^2)^{s_1+m_1+1}}{(z_1^2+v)^{r_1}}e^{-\frac{z_1^2}{2}}dz_1$$
$$=\int_0^{\infty}\frac{u_1^{s_1+m_1+1}}{(u_1^2+v)^{r_1}}e^{-\frac{u_1}{2}}\frac{1}{2\sqrt{u_1}}du_1+\int_0^{\infty}\frac{u_1^{s_1+m_1+1}}{(u_1^2+v)^{r_1}}e^{-\frac{u_1}{2}}\frac{1}{2\sqrt{u_1}}du_1$$
$$=\int_0^{\infty}\frac{u_1^{s_1+m_1+\frac{1}{2}}}{(u_1+v)^{r_1}}e^{-\frac{u_1}{2}}du_1.$$

Similarly,
$$\int_{-\infty}^{\infty}\frac{(z_2^2)^{s_2+m_2+1}}{(z_2^2+v)^{r_2}}e^{-\frac{z_2^2}{2}}dz_1=\int_0^{\infty}\frac{u_2^{s_2+m_2+\frac{1}{2}}}{(u_2+v)^{r_2}}e^{-\frac{u_2}{2}}du_2.$$

Therefore,

$$R_{m_1 m_2} = \int_0^\infty v^{q_1+q_2+\frac{\nu}{2}-1} e^{-\frac{v}{2}}$$
$$\times \left\{ \int_0^\infty \frac{u_1^{s_1+m_1+\frac{1}{2}}}{(u_1+v)^{r_1}} e^{-\frac{u_1}{2}} du_1 \right\} \left\{ \int_0^\infty \frac{u_2^{s_2+m_2+\frac{1}{2}}}{(u_2+v)^{r_2}} e^{-\frac{u_2}{2}} du_2 \right\} dv$$
$$= \int_0^\infty \int_0^\infty \int_0^\infty \frac{u_1^{s_1+m_1+\frac{1}{2}}}{(u_1+v)^{r_1}} \frac{u_2^{s_2+m_2+\frac{1}{2}}}{(u_2+v)^{r_2}} v^{q_1+q_2+\frac{\nu}{2}-1} e^{-\frac{u_1+u_2+v}{2}} du_1 du_2 dv.$$

Furthermore, using the change of variables

$$\begin{cases} t_1 := \dfrac{u_1}{u_1+u_2+v} \\ t_2 := \dfrac{u_2}{u_1+u_2+v} \\ t_3 := u_1+u_2+v \end{cases} \iff \begin{cases} u_1 = t_1 t_3 \\ u_2 = t_2 t_3 \\ v = t_3(1-t_1-t_2) \end{cases} \quad (B.9)$$

$$\implies du_1 du_2 dv = \text{abs}(|\mathbf{J}|) dt_1 dt_2 dt_3 = t_3^2 dt_1 dt_2 dt_3,$$

where $u_1, u_2, v > 0$, $t_3 > 0$, $(t_1, t_2) \in \mathcal{D} := \{(t_1, t_2) \; ; \; t_1, t_2 > 0, t_1 + t_2 < 1\}$, and $\text{abs}(x)$ is the absolute value of x, the Jacobian $|\mathbf{J}|$ is

$$|\mathbf{J}| := \begin{vmatrix} \frac{\partial u_1}{\partial t_1} & \frac{\partial u_1}{\partial t_2} & \frac{\partial u_1}{\partial t_3} \\ \frac{\partial u_2}{\partial t_1} & \frac{\partial u_2}{\partial t_2} & \frac{\partial u_2}{\partial t_3} \\ \frac{\partial v}{\partial t_1} & \frac{\partial v}{\partial t_2} & \frac{\partial v}{\partial t_3} \end{vmatrix} = \begin{vmatrix} t_3 & 0 & t_1 \\ 0 & t_3 & t_2 \\ -t_3 & -t_3 & 1-t_1-t_2 \end{vmatrix} = t_3^2.$$

From (B.9),

$$u_1 + v = t_3(1-t_2), \quad u_2 + v = t_3(1-t_1).$$

If we set

$$s_1 = \frac{p_1-1}{2}, \quad s_2 = \frac{p_2-1}{2},$$

then

$$R_{m_1 m_2} = \int_0^\infty \iint_{\mathcal{D}} \frac{(t_1 t_3)^{s_1+m_1+\frac{1}{2}}}{\{t_3(1-t_2)\}^{r_1}} \frac{(t_2 t_3)^{s_2+m_2+\frac{1}{2}}}{\{t_3(1-t_1)\}^{r_2}}$$

$$\times \{t_3(1-t_1-t_2)\}^{q_1+q_2+\frac{\nu}{2}-1} e^{-\frac{t_3}{2}} t_3^2 dt_1 dt_2 dt_3$$

$$= \int_0^\infty t_3^{s_1+m_1-r_1+\frac{1}{2}+s_2+m_2-r_2+\frac{1}{2}+q_1+q_2+\frac{\nu}{2}-1+2} e^{-\frac{t_3}{2}} dt_3$$

$$\times \iint_{\mathcal{D}} \frac{t_1^{s_1+m_1+\frac{1}{2}}}{(1-t_2)^{r_1}} \frac{t_2^{s_2+m_2+\frac{1}{2}}}{(1-t_1)^{r_2}} (1-t_1-t_2)^{q_1+q_2+\frac{\nu}{2}-1} dt_1 dt_2$$

$$= \int_0^\infty t_3^{m_1+m_2-r_1-r_2+\frac{p_1+p_2+\nu}{2}+q_1+q_2+2-1} e^{-\frac{t_3}{2}} dt_3$$

$$\times \iint_{\mathcal{D}} \frac{1}{(1-t_1)^{r_2}} \frac{1}{(1-t_2)^{r_1}} t_1^{m_1+\frac{p_1}{2}} t_2^{m_2+\frac{p_2}{2}} (1-t_1-t_2)^{q_1+q_2+\frac{\nu}{2}-1} dt_1 dt_2$$

$$= 2^{m_1+m_2+q_1+q_2-r_1-r_2+\frac{p_1+p_2+\nu+4}{2}}$$

$$\times \Gamma\left(m_1+m_2+q_1+q_2-r_1-r_2+\frac{p_1+p_2+\nu+4}{2}\right)$$

$$\times \mathcal{B}_{r_1 r_2}\left(m_1+\frac{p_1+2}{2}, m_2+\frac{p_2+2}{2}, q_1+q_2+\frac{\nu}{2}\right). \quad (B.10)$$

If we substitute this result into (B.8), then we get

$$E\left(\frac{Z_1^{p_1} V^{q_1}}{(Z_1^2+V)^{r_1}} \frac{Z_2^{p_2} V^{q_2}}{(Z_2^2+V)^{r_2}}\right)$$

$$= \tau_1 \tau_2 2^{q_1+q_2-r_1-r_2+\frac{p_1+p_2}{2}-1} \sum_{m_1=0}^\infty \sum_{m_2=0}^\infty \frac{\Gamma\left(m_1+m_2+q_1+q_2-r_1-r_2+\frac{p_1+p_2+\nu}{2}+\right.}{\Gamma\left(m_1+\frac{3}{2}\right)\Gamma\left(m_2+\frac{3}{2}\right)\Gamma\left(\frac{\nu}{2}\right)}$$

$$\times \mathcal{B}_{r_1 r_2}\left(m_1+\frac{p_1+2}{2}, m_2+\frac{p_2+2}{2}, q_1+q_2+\frac{\nu}{2}\right) p(m_1; \tau_1^2/2) p(m_2; \tau_2^2/2).$$

The proof is complete. □

Appendix C
Data Sets

C.1 Data Generating Model and Data Sets

The data sets shown in Tables C.1 and C.2 are based on those in Hoerl (1962) [1], which have frequently been used for examining multicollinearity. These data sets are generated by the following model:

$$\begin{aligned} y_i &= \theta_0 + \theta_1 z_{i1} + \theta_2 z_{i2} + \theta_3 z_{i3} + \epsilon_i \\ &= 10 + 2z_{i1} + 3z_{i2} + 5z_{i3} + \epsilon_i, \quad i = 1, \cdots, 10 (= n). \end{aligned}$$

The matrix-vector representation is

$$\boldsymbol{y} = \mathbf{Z}\boldsymbol{\theta} + \boldsymbol{\epsilon} = \theta_0 \mathbf{1} + \mathbf{Z}_p \boldsymbol{\theta}_p + \boldsymbol{\epsilon},$$

where

$$\boldsymbol{y} = \begin{bmatrix} y_1 \\ \vdots \\ y_{10} \end{bmatrix}, \quad \mathbf{Z} := [\mathbf{1}, \mathbf{Z}_p] := \begin{bmatrix} 1 & z_{11} & z_{12} & z_{13} \\ \vdots & \vdots & \vdots & \vdots \\ 1 & z_{10\ 1} & z_{10\ 2} & z_{10\ 3} \end{bmatrix},$$

$$\boldsymbol{\theta} = \begin{bmatrix} \theta_0 \\ \boldsymbol{\theta}_p \end{bmatrix} = \begin{bmatrix} 10 \\ 2 \\ 3 \\ 5 \end{bmatrix}, \quad \boldsymbol{\epsilon} := \begin{bmatrix} \epsilon_1 \\ \vdots \\ \epsilon_{10} \end{bmatrix},$$

and

$$\mathbf{Z}_p := \begin{bmatrix} z_{11} & z_{12} & z_{13} \\ \vdots & \vdots & \vdots \\ z_{10\ 1} & z_{10\ 2} & z_{10\ 3} \end{bmatrix}, \quad \boldsymbol{\theta}_p = \begin{bmatrix} \theta_1 \\ \theta_2 \\ \theta_3 \end{bmatrix} = \begin{bmatrix} 2 \\ 3 \\ 5 \end{bmatrix}, \quad p = 3.$$

---

[1] These data sets differ from the originals in that the response variables, y_i, are newly generated with random errors in order to increase the accuracy of the simulations.

C DATA SETS

Table C.1: Data Set 1					Table C.2: Data Set 2				
No. i	z_{i1}	z_{i2}	z_{i3}	y_i	No. i	z_{i1}	z_{i2}	z_{i3}	y_i
1	1.1	1.1	1.2	22.7	1	1.1	1.1	1.2	20.6
2	1.2	1.4	1.2	21.7	2	1.4	1.5	1.1	23.7
3	1.3	1.2	1.6	24.2	3	1.7	1.8	2.0	28.1
4	1.5	1.4	1.5	25.9	4	1.7	1.7	1.8	26.1
5	1.6	1.8	1.6	26.7	5	1.8	1.9	1.8	28.1
6	1.9	1.7	1.9	27.9	6	1.8	1.8	1.9	29.4
7	1.9	2.0	2.2	31.5	7	1.9	1.8	2.0	29.9
8	2.3	2.2	2.1	33.5	8	2.0	2.1	2.1	31.5
9	2.4	2.5	2.4	35.2	9	2.3	2.4	2.5	33.9
10	2.5	2.2	2.3	33.0	10	2.5	2.5	2.4	33.7

We consider several statistics with respect to the usual model (M_Z), the standardized model (M_X), and the canonical form (M_A) defined in Chapter 1.

C.2 (Data Set 1)

A matrix of scatter plots of the data set is given in Figure C.1.

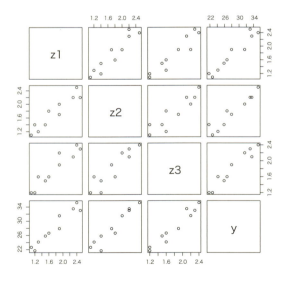

Figure C.1: Pairwise Plot: Data Set 1

C.2.1 Usual Model

The matrix of the explanatory variables and the vector of response variables are

$$\mathbf{Z} = \begin{bmatrix} 1 & 1.1 & 1.1 & 1.2 \\ 1 & 1.2 & 1.4 & 1.2 \\ 1 & 1.3 & 1.2 & 1.6 \\ 1 & 1.5 & 1.4 & 1.5 \\ 1 & 1.6 & 1.8 & 1.6 \\ 1 & 1.9 & 1.7 & 1.9 \\ 1 & 1.9 & 2.0 & 2.2 \\ 1 & 2.3 & 2.2 & 2.1 \\ 1 & 2.4 & 2.5 & 2.4 \\ 1 & 2.5 & 2.2 & 2.3 \end{bmatrix}, \quad \mathbf{y} = \begin{bmatrix} 20.6 \\ 23.7 \\ 28.1 \\ 26.1 \\ 28.1 \\ 29.4 \\ 29.9 \\ 31.5 \\ 33.9 \\ 33.7 \end{bmatrix}.$$

The information matrix $\mathbf{Z}'\mathbf{Z}$ and its inverse matrix are

$$\mathbf{Z}'\mathbf{Z} = \begin{bmatrix} 10.0 & 17.7 & 17.5 & 18.0 \\ 17.7 & 33.67 & 33.02 & 33.78 \\ 17.5 & 33.02 & 32.63 & 33.21 \\ 18.0 & 33.78 & 33.21 & 34.16 \end{bmatrix},$$

$$(\mathbf{Z}'\mathbf{Z})^{-1} = \begin{bmatrix} 2.19 & 1.27 & -0.67 & -1.76 \\ 1.27 & 6.55 & -3.42 & -3.82 \\ -0.67 & -3.42 & 4.70 & -0.83 \\ -1.76 & -3.82 & -0.83 & 5.54 \end{bmatrix}.$$

The OLS estimate $\widehat{\boldsymbol{\theta}}$ and its variance-covariance matrix are

$$\widehat{\boldsymbol{\theta}} = (\mathbf{Z}'\mathbf{Z})^{-1}\mathbf{Z}'\mathbf{y} = \begin{bmatrix} 9.7 \\ 2.2 \\ 3.1 \\ 5.1 \end{bmatrix},$$

$$V(\widehat{\boldsymbol{\theta}}) = \sigma^2 (\mathbf{Z}'\mathbf{Z})^{-1} = \begin{bmatrix} 2.19 & 1.27 & -0.67 & -1.76 \\ 1.27 & 6.55 & -3.42 & -3.82 \\ -0.67 & -3.42 & 4.70 & -0.83 \\ -1.76 & -3.82 & -0.83 & 5.54 \end{bmatrix}.$$

The estimate for σ^2 is $\widehat{\sigma}^2 = \|y - \mathbf{Z}\widehat{\boldsymbol{\theta}}\|^2/(n-p-1) = 1.0074$.

C.2.2 Standardized Model

The transformation of the standardization $\mathbf{T}$ and its inverse $\mathbf{T}^{-1}$ are

$$\mathbf{T} = \begin{bmatrix} 1 & \boldsymbol{m}'_p \\ \mathbf{0} & \mathbf{S}_p \end{bmatrix} = \begin{bmatrix} 1 & 1.77 & 1.75 & 1.8 \\ 0 & \sqrt{2.34} & 0 & 0 \\ 0 & 0 & \sqrt{2.01} & 0 \\ 0 & 0 & 0 & \sqrt{1.76} \end{bmatrix} = \begin{bmatrix} 1 & 1.77 & 1.75 & 1.8 \\ 0 & 1.53 & 0 & 0 \\ 0 & 0 & 1.42 & 0 \\ 0 & 0 & 0 & 1.33 \end{bmatrix},$$

$$\mathbf{T}^{-1} = \begin{bmatrix} 1 & -\boldsymbol{m}'_p\mathbf{S}_p^{-1} \\ \mathbf{0} & \mathbf{S}_p^{-1} \end{bmatrix} = \begin{bmatrix} 1 & -1.16 & -1.24 & -1.36 \\ 0 & 0.65 & 0 & 0 \\ 0 & 0 & 0.71 & 0 \\ 0 & 0 & 0 & 0.75 \end{bmatrix}.$$

The standardized matrix of $\mathbf{Z}$ is

$$\mathbf{X} = \mathbf{Z}\mathbf{T}^{-1} = \begin{bmatrix} 1 & 1.1 & 1.1 & 1.2 \\ 1 & 1.2 & 1.4 & 1.2 \\ 1 & 1.3 & 1.2 & 1.6 \\ 1 & 1.5 & 1.4 & 1.5 \\ 1 & 1.6 & 1.8 & 1.6 \\ 1 & 1.9 & 1.7 & 1.9 \\ 1 & 1.9 & 2.0 & 2.2 \\ 1 & 2.3 & 2.2 & 2.1 \\ 1 & 2.4 & 2.5 & 2.4 \\ 1 & 2.5 & 2.2 & 2.3 \end{bmatrix} \begin{bmatrix} 1 & -1.16 & -1.24 & -1.36 \\ 0 & 0.65 & 0 & 0 \\ 0 & 0 & 0.71 & 0 \\ 0 & 0 & 0 & 0.75 \end{bmatrix}$$

$$= \begin{bmatrix} 1 & -0.44 & -0.46 & -0.45 \\ 1 & -0.37 & -0.25 & -0.45 \\ 1 & -0.31 & -0.39 & -0.15 \\ 1 & -0.18 & -0.25 & -0.23 \\ 1 & -0.11 & 0.04 & -0.15 \\ 1 & 0.08 & -0.04 & 0.08 \\ 1 & 0.08 & 0.18 & 0.30 \\ 1 & 0.35 & 0.32 & 0.23 \\ 1 & 0.41 & 0.53 & 0.45 \\ 1 & 0.48 & 0.32 & 0.38 \end{bmatrix}.$$

The vector of regression coefficient $\boldsymbol{\beta}$ and its OLS estimate $\widehat{\boldsymbol{\beta}}$ are

$$\boldsymbol{\beta} = \mathbf{T}\boldsymbol{\theta} = \begin{bmatrix} 1 & 1.77 & 1.75 & 1.8 \\ 0 & 1.53 & 0 & 0 \\ 0 & 0 & 1.42 & 0 \\ 0 & 0 & 0 & 1.33 \end{bmatrix} \begin{bmatrix} 10 \\ 2 \\ 3 \\ 5 \end{bmatrix} = \begin{bmatrix} 27.8 \\ 3.1 \\ 4.2 \\ 6.6 \end{bmatrix},$$

$$\widehat{\boldsymbol{\beta}} = (\mathbf{X}'\mathbf{X})^{-1}\mathbf{X}'\boldsymbol{y} = \begin{bmatrix} 28.2 \\ 3.4 \\ 4.4 \\ 6.7 \end{bmatrix}.$$

The correlation form $\mathbf{X}'_p\mathbf{X}_p$ and its inverse $(\mathbf{X}'_p\mathbf{X}_p)^{-1}$ are

$$\mathbf{X}'_p\mathbf{X}_p = \begin{bmatrix} 1 & 0.94 & 0.95 \\ 0.94 & 1 & 0.91 \\ 0.95 & 0.91 & 1 \end{bmatrix}, \quad (\mathbf{X}'_p\mathbf{X}_p)^{-1} = \begin{bmatrix} 15.34 & -7.42 & -7.76 \\ -7.42 & 9.42 & -1.56 \\ -7.76 & -1.56 & 9.76 \end{bmatrix}.$$

The variance-covariance matrix of the OLS estimator $\widehat{\boldsymbol{\beta}}_p$ is

$$V(\widehat{\boldsymbol{\beta}}_p) = \sigma^2(\mathbf{X}'_p\mathbf{X}_p)^{-1} = \begin{bmatrix} 15.34 & -7.42 & -7.76 \\ -7.42 & 9.42 & -1.56 \\ -7.76 & -1.56 & 9.76 \end{bmatrix}.$$

Hence, its trace (the total variance) is

$$traceV(\widehat{\boldsymbol{\beta}}_p) = trace\,\sigma^2(\mathbf{X}'_p\mathbf{X}_p)^{-1} = 15.34 + 9.42 + 9.76 = 34.52,$$

and the variance inflation factors VIF_i $(i = 1, 2, 3)$ are

$$VIF_1 = 15.34, \quad VIF_2 = 9.42, \quad VIF_3 = 9.76.$$

C.2.3 Canonical Form

The matrix of principal components is

$$\mathbf{A} = \mathbf{X}\boldsymbol{\Gamma} = \begin{bmatrix} 1 & -0.44 & -0.46 & -0.45 \\ 1 & -0.37 & -0.25 & -0.45 \\ 1 & -0.31 & -0.39 & -0.15 \\ 1 & -0.18 & -0.25 & -0.23 \\ 1 & -0.11 & 0.04 & -0.15 \\ 1 & 0.08 & -0.04 & 0.08 \\ 1 & 0.08 & 0.18 & 0.30 \\ 1 & 0.35 & 0.32 & 0.23 \\ 1 & 0.41 & 0.53 & 0.45 \\ 1 & 0.48 & 0.32 & 0.38 \end{bmatrix} \begin{bmatrix} 1 & 0 & 0 & 0 \\ 0 & -0.58 & 0.02 & 0.81 \\ 0 & -0.57 & -0.72 & -0.39 \\ 0 & -0.58 & 0.7 & -0.43 \end{bmatrix}$$

$$= \begin{bmatrix} 1 & 0.78 & 0.01 & 0.02 \\ 1 & 0.62 & -0.15 & -0.01 \\ 1 & 0.49 & 0.17 & -0.03 \\ 1 & 0.37 & 0.02 & 0.05 \\ 1 & 0.13 & -0.13 & -0.04 \\ 1 & -0.07 & 0.08 & 0.05 \\ 1 & -0.32 & 0.09 & -0.13 \\ 1 & -0.51 & -0.06 & 0.06 \\ 1 & -0.8 & -0.06 & -0.07 \\ 1 & -0.68 & 0.04 & 0.10 \end{bmatrix},$$

where $\boldsymbol{\Gamma}$ is an orthogonal transformation. The scree plot of principal components is given in Figure C.2.

The vector of regression coefficients $\boldsymbol{\alpha}$ and its OLS estimate are

$$\boldsymbol{\alpha} = \boldsymbol{\Gamma}'\boldsymbol{\beta} = \begin{bmatrix} 1 & 0 & 0 & 0 \\ 0 & -0.58 & 0.02 & 0.81 \\ 0 & -0.57 & -0.72 & -0.39 \\ 0 & -0.58 & 0.70 & -0.43 \end{bmatrix}' \begin{bmatrix} 27.79 \\ 3.06 \\ 4.25 \\ 6.63 \end{bmatrix} = \begin{bmatrix} 27.8 \\ -8.0 \\ 1.6 \\ -2.0 \end{bmatrix},$$

$$\widehat{\boldsymbol{\alpha}} = (\mathbf{A}'\mathbf{A})^{-1}\mathbf{A}'\boldsymbol{y} = \begin{bmatrix} 28.2 \\ -8.4 \\ 1.6 \\ -1.9 \end{bmatrix}.$$

The information matrix $\mathbf{A}'_p\mathbf{A}_p$ and its inverse matrix are

$$\mathbf{A}'_p\mathbf{A}_p = \boldsymbol{\Lambda}_p = \mathrm{diag}(\lambda_1, \lambda_2, \lambda_3) = \mathrm{diag}(2.87, 0.09, 0.04),$$
$$(\mathbf{A}'_p\mathbf{A}_p)^{-1} = \boldsymbol{\Lambda}_p^{-1} = \mathrm{diag}(\lambda_1^{-1}, \lambda_2^{-1}, \lambda_3^{-1}) = \mathrm{diag}(0.35, 11.14, 23.03),$$

where $\{\lambda_1, \lambda_2, \lambda_3\}$ are eigenvalues of the matrix $\mathbf{X}'_p\mathbf{X}_p$.

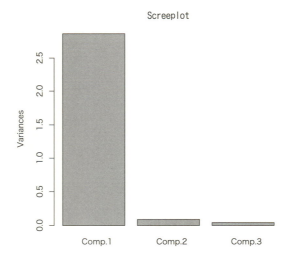

Figure C.2: Scree plot for the correlation form: Data Set 1

The following tables list several statistics for the comparisons among the

MSE criteria of the shrinkage regression estimators:

i	λ_i	α_i	α_i^2	k_i^*	τ_i	τ_i^2	δ_i
1	2.87	-8.04	64.61	0.02	-13.61	185.23	92.62
2	0.09	1.63	2.67	0.37	0.49	0.24	0.12
3	0.04	-2.03	4.12	0.24	-0.42	0.18	0.09
i	λ_i	$\widehat{\alpha}_i$	$\widehat{\alpha}_i^2$	$\widehat{k}_i^*$	$\widehat{\tau}_i$	$\widehat{\tau}_i^2$	$\widehat{\delta}_i$
1	2.87	-8.4	70.63	0.01	-14.18	200.99	100.5
2	0.09	1.57	2.46	0.41	0.47	0.22	0.11
3	0.04	-1.87	3.52	0.29	-0.39	0.15	0.08

where

$$k_i^* := \frac{\sigma^2}{\alpha_i^2}, \quad \tau_i := \frac{\alpha_i}{\sigma/\sqrt{\lambda_i}}, \quad \delta_i := \tau_i^2/2,$$

$$\widehat{k}_i^* := \frac{\widehat{\sigma}^2}{\widehat{\alpha}_i^2}, \quad \widehat{\tau}_i := \frac{\widehat{\alpha}_i}{\widehat{\sigma}/\sqrt{\lambda_i}}, \quad \widehat{\delta}_i := \widehat{\tau}_i^2/2.$$

C.3 (Data Set 2)

A matrix of scatter plots of the data set is given in Figure C.3.

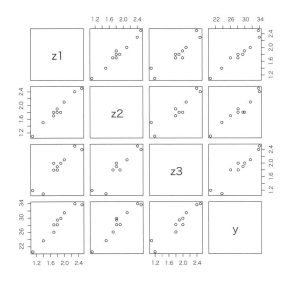

Figure C.3: Pairwise Plot: Data Set 2

C.3.1 Usual Model

The matrix of the explanatory variables and the vector of response variables are

$$\mathbf{Z} = \begin{bmatrix} 1 & 1.1 & 1.1 & 1.2 \\ 1 & 1.4 & 1.5 & 1.1 \\ 1 & 1.7 & 1.8 & 2.0 \\ 1 & 1.7 & 1.7 & 1.8 \\ 1 & 1.8 & 1.9 & 1.8 \\ 1 & 1.8 & 1.8 & 1.9 \\ 1 & 1.9 & 1.8 & 2.0 \\ 1 & 2.0 & 2.1 & 2.1 \\ 1 & 2.3 & 2.4 & 2.5 \\ 1 & 2.5 & 2.5 & 2.4 \end{bmatrix}, \quad \mathbf{y} = \begin{bmatrix} 20.6 \\ 23.7 \\ 28.1 \\ 26.1 \\ 28.1 \\ 29.4 \\ 29.9 \\ 31.5 \\ 33.9 \\ 33.7 \end{bmatrix}.$$

The information matrix $\mathbf{Z'Z}$ and its inverse matrix are

$$\mathbf{Z'Z} = \begin{bmatrix} 10 & 18.2 & 18.6 & 18.8 \\ 18.2 & 34.58 & 35.31 & 35.73 \\ 18.6 & 35.31 & 36.1 & 36.48 \\ 18.8 & 35.73 & 36.48 & 37.16 \end{bmatrix},$$

$$(\mathbf{Z'Z})^{-1} = \begin{bmatrix} 2.41 & -0.44 & -0.85 & 0.04 \\ -0.44 & 28.79 & -23.2 & -4.69 \\ -0.85 & -23.2 & 22.76 & 0.38 \\ 0.04 & -4.69 & 0.38 & 4.14 \end{bmatrix}.$$

The OLS estimate $\widehat{\boldsymbol{\theta}}$ and its variance-covariance matrix are

$$\widehat{\boldsymbol{\theta}} = (\mathbf{Z'Z})^{-1}\mathbf{Z'y} = \begin{bmatrix} 9.8 \\ 3.5 \\ 3.2 \\ 3.4 \end{bmatrix},$$

$$V(\widehat{\boldsymbol{\theta}}) = \sigma^2 (\mathbf{Z'Z})^{-1} = \begin{bmatrix} 2.41 & -0.44 & -0.85 & 0.04 \\ -0.44 & 28.79 & -23.20 & -4.69 \\ -0.85 & -23.2 & 22.76 & 0.38 \\ 0.04 & -4.69 & 0.38 & 4.14 \end{bmatrix}.$$

The estimate for σ^2 is $\widehat{\sigma}^2 = \|\mathbf{y} - \mathbf{Z}\widehat{\boldsymbol{\theta}}\|^2/(n-p-1) = 1.0002$.

C.3.2 Standardized Model

The transformation of the standardization $\mathbf{T}$ and its inverse $\mathbf{T}^{-1}$ are

$$\mathbf{T} = \begin{bmatrix} 1 & \mathbf{m}'_p \\ \mathbf{0} & \mathbf{S}_p \end{bmatrix} = \begin{bmatrix} 1 & 1.82 & 1.86 & 1.88 \\ 0 & \sqrt{1.46} & 0 & 0 \\ 0 & 0 & \sqrt{1.5} & 0 \\ 0 & 0 & 0 & \sqrt{1.82} \end{bmatrix} = \begin{bmatrix} 1 & 1.82 & 1.86 & 1.88 \\ 0 & 1.21 & 0 & 0 \\ 0 & 0 & 1.23 & 0 \\ 0 & 0 & 0 & 1.35 \end{bmatrix},$$

$$\mathbf{T}^{-1} = \begin{bmatrix} 1 & -\mathbf{m}'_p \mathbf{S}_p^{-1} \\ \mathbf{0} & \mathbf{S}_p^{-1} \end{bmatrix} = \begin{bmatrix} 1 & -1.51 & -1.52 & -1.4 \\ 0 & 0.83 & 0 & 0 \\ 0 & 0 & 0.82 & 0 \\ 0 & 0 & 0 & 0.74 \end{bmatrix}.$$

The standardized matrix of $\mathbf{Z}$ is

$$\mathbf{X} = \mathbf{ZT}^{-1} = \begin{bmatrix} 1 & 1.1 & 1.1 & 1.2 \\ 1 & 1.4 & 1.5 & 1.1 \\ 1 & 1.7 & 1.8 & 2.0 \\ 1 & 1.7 & 1.7 & 1.8 \\ 1 & 1.8 & 1.9 & 1.8 \\ 1 & 1.8 & 1.8 & 1.9 \\ 1 & 1.9 & 1.8 & 2.0 \\ 1 & 2.0 & 2.1 & 2.1 \\ 1 & 2.3 & 2.4 & 2.5 \\ 1 & 2.5 & 2.5 & 2.4 \end{bmatrix} \begin{bmatrix} 1 & -1.51 & -1.52 & -1.4 \\ 0 & 0.83 & 0 & 0 \\ 0 & 0 & 0.82 & 0 \\ 0 & 0 & 0 & 0.74 \end{bmatrix}$$

$$= \begin{bmatrix} 1 & -0.6 & -0.62 & -0.50 \\ 1 & -0.35 & -0.29 & -0.58 \\ 1 & -0.10 & -0.05 & 0.09 \\ 1 & -0.10 & -0.13 & -0.06 \\ 1 & -0.02 & 0.03 & -0.06 \\ 1 & -0.02 & -0.05 & 0.01 \\ 1 & 0.07 & -0.05 & 0.09 \\ 1 & 0.15 & 0.20 & 0.16 \\ 1 & 0.40 & 0.44 & 0.46 \\ 1 & 0.56 & 0.52 & 0.39 \end{bmatrix}.$$

The vector of regression coefficient $\boldsymbol{\beta}$ and its OLS estimate $\widehat{\boldsymbol{\beta}}$ are

$$\boldsymbol{\beta} = \mathbf{T}\boldsymbol{\theta} = \begin{bmatrix} 1 & 1.82 & 1.86 & 1.88 \\ 0 & 1.21 & 0 & 0 \\ 0 & 0 & 1.23 & 0 \\ 0 & 0 & 0 & 1.35 \end{bmatrix} \begin{bmatrix} 10 \\ 2 \\ 3 \\ 5 \end{bmatrix} = \begin{bmatrix} 28.6 \\ 2.4 \\ 3.7 \\ 6.7 \end{bmatrix},$$

$$\widehat{\boldsymbol{\beta}} = (\mathbf{X}'\mathbf{X})^{-1}\mathbf{X}'\boldsymbol{y} = \begin{bmatrix} 28.5 \\ 4.2 \\ 3.9 \\ 4.6 \end{bmatrix}.$$

The correlation form $\mathbf{X}'_p \mathbf{X}_p$ and its inverse $(\mathbf{X}'_p \mathbf{X}_p)^{-1}$ are given by

$$\mathbf{X}'_p \mathbf{X}_p = \begin{bmatrix} 1 & 0.99 & 0.93 \\ 0.99 & 1 & 0.91 \\ 0.93 & 0.91 & 1 \end{bmatrix}, \quad (\mathbf{X}'_p \mathbf{X}_p)^{-1} = \begin{bmatrix} 41.93 & -34.32 & -7.63 \\ -34.32 & 34.24 & 0.64 \\ -7.63 & 0.64 & 7.53 \end{bmatrix}.$$

The variance-covariance matrix of the OLS estimator $\widehat{\boldsymbol{\beta}}_p$ is

$$V(\widehat{\boldsymbol{\beta}}_p) = \sigma^2 (\mathbf{X}'_p \mathbf{X}_p)^{-1} = \begin{bmatrix} 41.93 & -34.32 & -7.63 \\ -34.32 & 34.24 & 0.64 \\ -7.63 & 0.64 & 7.53 \end{bmatrix}.$$

116   C DATA SETS

Hence, its trace (the total variance) is

$$\text{trace} V(\widehat{\boldsymbol{\beta}}_p) = \text{trace } \sigma^2 (\mathbf{X}'_p \mathbf{X}_p)^{-1} = 41.93 + 34.24 + 7.53 = 83.69,$$

and the variance inflation factors VIF_i $(i = 1, 2, 3)$ are

$$\text{VIF}_1 = 41.93, \quad \text{VIF}_2 = 34.24, \quad \text{VIF}_3 = 7.53.$$

C.3.3   Canonical Form

The matrix of principal components of $\mathbf{X}$ is

$$\mathbf{A} = \mathbf{X}\boldsymbol{\Gamma} = \begin{bmatrix} 1 & -0.6 & -0.62 & -0.50 \\ 1 & -0.35 & -0.29 & -0.58 \\ 1 & -0.1 & -0.05 & 0.09 \\ 1 & -0.1 & -0.13 & -0.06 \\ 1 & -0.02 & 0.03 & -0.06 \\ 1 & -0.02 & -0.05 & 0.01 \\ 1 & 0.07 & -0.05 & 0.09 \\ 1 & 0.15 & 0.20 & 0.16 \\ 1 & 0.4 & 0.44 & 0.46 \\ 1 & 0.56 & 0.52 & 0.39 \end{bmatrix} \begin{bmatrix} 1 & 0 & 0 & 0 \\ 0 & 0.58 & -0.32 & 0.75 \\ 0 & 0.58 & -0.48 & -0.66 \\ 0 & 0.57 & 0.82 & -0.09 \end{bmatrix}$$

$$= \begin{bmatrix} 1 & -0.99 & 0.08 & 0.01 \\ 1 & -0.7 & -0.22 & -0.01 \\ 1 & -0.04 & 0.13 & -0.05 \\ 1 & -0.17 & 0.05 & 0.02 \\ 1 & -0.02 & -0.06 & -0.03 \\ 1 & -0.03 & 0.04 & 0.02 \\ 1 & 0.06 & 0.07 & 0.07 \\ 1 & 0.29 & -0.01 & -0.03 \\ 1 & 0.75 & 0.04 & -0.04 \\ 1 & 0.85 & -0.12 & 0.04 \end{bmatrix},$$

where $\boldsymbol{\Gamma}$ is a orthogonal transformation. The scree plot of principal components is given in Figure C.4.

The vector of regression coefficients $\boldsymbol{\alpha}$ and its OLS estimate are

$$\boldsymbol{\alpha} = \boldsymbol{\Gamma}' \boldsymbol{\beta} = \begin{bmatrix} 1 & 0 & 0 & 0 \\ 0 & 0.58 & -0.32 & 0.75 \\ 0 & 0.58 & -0.48 & -0.66 \\ 0 & 0.57 & 0.82 & -0.09 \end{bmatrix}' \begin{bmatrix} 28.62 \\ 2.41 \\ 3.68 \\ 6.74 \end{bmatrix} = \begin{bmatrix} 28.6 \\ 7.4 \\ 3 \\ -1.3 \end{bmatrix},$$

$$\widehat{\boldsymbol{\alpha}} = (\mathbf{A}'\mathbf{A})^{-1} \mathbf{A}' \mathbf{y} = \begin{bmatrix} 28.5 \\ 7.3 \\ 0.6 \\ 0.2 \end{bmatrix}.$$

The information matrix $\mathbf{A}'_p\mathbf{A}_p$ and its inverse matrix are

$$\mathbf{A}'_p\mathbf{A}_p = \mathbf{\Lambda}_p = \mathrm{diag}(\lambda_1, \lambda_2, \lambda_3) = \mathrm{diag}(2.89, 0.10, 0.01),$$
$$(\mathbf{A}'_p\mathbf{A}_p)^{-1} = \mathbf{\Lambda}_p^{-1} = \mathrm{diag}(\lambda_1^{-1}, \lambda_2^{-1}, \lambda_3^{-1}) = \mathrm{diag}(0.35, 10.15, 73.19),$$

where $\{\lambda_1, \lambda_2, \lambda_3\}$ are eigenvalues of the matrix $\mathbf{X}'_p\mathbf{X}_p$.

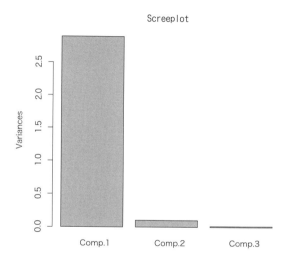

Figure C.4: Scree plot for the correlation form: Data Set 2

The following tables list several statistics for comparisons among the MSE criteria of shrinkage regression estimators:

i	λ_i	α_i	α_i^2	k_i^*	τ_i	τ_i^2	δ_i
1	2.89	7.37	54.37	0.02	12.53	157.01	78.5
2	0.10	2.97	8.83	0.11	0.93	0.87	0.43
3	0.01	-1.25	1.57	0.64	-0.15	0.02	0.01

i	λ_i	$\widehat{\alpha}_i$	$\widehat{\alpha}_i^2$	$\widehat{k}_i^*$	$\widehat{\tau}_i$	$\widehat{\tau}_i^2$	$\widehat{\delta}_i$
1	2.89	7.32	53.6	0.02	12.44	154.77	77.38
2	0.10	0.56	0.31	3.22	0.17	0.03	0.02
3	0.01	0.17	0.03	35.16	0.02	0.00	0.00

where

$$k_i^* := \frac{\sigma^2}{\alpha_i^2}, \quad \tau_i := \frac{\alpha_i}{\sigma/\sqrt{\lambda_i}}, \quad \delta_i := \tau_i^2/2,$$

$$\widehat{k_i^*} := \frac{\widehat{\sigma}^2}{\widehat{\alpha}_i^2}, \quad \widehat{\tau}_i := \frac{\widehat{\alpha}_i}{\widehat{\sigma}/\sqrt{\lambda_i}}, \quad \widehat{\delta}_i := \widehat{\tau}_i^2/2.$$

Bibliography

[1] D. M. Allen. Mean square error of prediction as a criterion for selecting variables. *Technometrics*, Vol. 13, pp. 469–475, 1971.

[2] D. M. Allen. The relationship between variable selection and data augmentation and a method for prediction. *Technometrics*, Vol. 16, pp. 125–127, 1974.

[3] K. F. Baldwin and A. E. Hoerl. Bounds on minimum mean squared error in ridge regression. *Communications in Statistics. Theory and Method*, Vol. A7, pp. 1209–1218, 1978.

[4] M. R. Baye and D. F. Parker. Combining ridge and principal component regression: A money demand illustration. *Communication in Statistics. Theory and Method*, Vol. A13, pp. 197–205, 1984.

[5] D. A. Belsley, E. Kuh, and R. E. Welsch. *Regression Diagnostics*. John Wiley and Sons, Inc., 1980.

[6] P. Billingsley. *Probability and Measure*. John Wiley and Sons, Inc., third edition, 1995.

[7] C. M. Bishop. *Pattern Recognition and Machine Learning*. Springer-Verlag, 2006.

[8] P. J. Brown. Centering and scaling in ridge regression. *Technometrics*, Vol. 19, pp. 35–36, 1977.

[9] A. C. Davison and D. V. Hinkley. *Bootstrap Methods and Their Application*. Cambridge University Press, 1997.

[10] A. P. Dempster, M. Schatzoff, and N. Wermuth. A simulation study of alternatives to ordinary least squares. *Journal of the American Statistical Association*, Vol. 72, pp. 77–106, 1977.

[11] N. R. Draper and H. Smith. *Applied Regression Analysis*. John Wiley and Sons, Inc., second edition, 1981.

[12] T. D. Dwivedi, V. K. Srivastava, and R. L. Hall. Finite sample properties of ridge estimators. *Technometrics*, Vol. 22, pp. 205–212, 1980.

[13] B. Efron. Bootstrap methods: another look at the jackknife. *Annals of Statistics*, Vol. 7, pp. 1–26, 1979.

[14] B. Efron and R. J. Tibshirani. *An Introduction to the Bootstrap*. Chapman and Hall, 1993.

[15] L. Firinguetti. A generalized ridge regression estimator and its finite sample properties. *Communications in Statistics. Theory and Method*, Vol. 28, No. 5, pp. 1217–1229, 1999.

[16] D. G. Gibbons. A simulation study of regression estimators. *Journal of the American Statistical Association*, Vol. 76, pp. 131–139, 1981.

[17] R. F. Gunst. Regression analysis with multicollinear predictor variables: Definition, detection, and effects. *Communication in Statistics. Theory and Method*, Vol. 12, pp. 2217–2260, 1983.

[18] L. Györfi, M. Kohler, A. Krzyżak, and H. Walk. *A Distibution-Free Theory of Nonparametric Regression*. Springer-Verlag, 2002.

[19] T. Hastie, R. Tibshirani, and J. Friedman. *The Elements of Statistical Learning: Data Mining, Inference, and Prediction*. Springer-Verlag, second edition, 2009.

[20] W. J. Hemmerle and M. B. Carey. Some properties of generalized ridge estimators. *Communications in Statistics. Computation and Simulation*, Vol. 12, pp. 239–256, 1983.

[21] A. E. Hoerl. Optimum solution of many variables equations. *Chemical Engineering Progress*, Vol. 55, pp. 67–78, 1959.

[22] A. E. Hoerl. Application of ridge analysis to regression problems. *Chemical Engineering Progress*, Vol. 58, pp. 54–59, 1962.

[23] A. E. Hoerl and R. W. Kennard. Ridge regression: biased estimation for nonorthogonal problems. *Technometrics*, Vol. 12, pp. 55–67, 1970a.

[24] A. E. Hoerl and R. W. Kennard. Ridge regression: application to nonorthogonal problems. *Technometrics*, Vol. 12, pp. 69–82, 1970b.

[25] A. E. Hoerl, R. W. Kennard, and K. F. Baldwin. Ridge regression: some simulations. *Communications in Statistics*, Vol. 4, pp. 105–123, 1975.

[26] N. Inagaki. Two errors in statistical model fitting. *Annals of the Institute of Statistical Mathematics*, pp. 131–152, 1977.

[27] T. Inoue. Density function and relative efficiency of the modified generalized ridge regression estimators. *Journal of the Japan Statistical Society*, Vol. 29, pp. 39–54, 1999.

[28] W. James and C. Stein. Estimation with quadratic loss. In *Proceedings of Fourth Berkeley Symposium on Mathematical Statistics and Probability*, Vol. I, pp. 361–379. University of California Press, Berkeley, 1961.

[29] M. Jimichi. Mean squared error criteria of feasible generalized ridge regression estimator. *International Review of Business*, No. 4, pp. 62–74, 1999.

[30] M. Jimichi. *Improvement of Regression Estimators by Shrinkage under Multicollinearity and Its Feasibility.* PhD thesis, Osaka University, 2005.

[31] M. Jimichi. Numerical evaluations for exact moments of feasible generalized ridge regression estimator. *K.G. Studies in Computer Science*, Vol. 21, pp. 3–22, 2007 (in Japanese).

[32] M. Jimichi. Exact moments of feasible generalized ridge regression estimator and numerical evaluations. *Journal of the Japanese Society of Computational Statistics*, Vol. 21, pp. 1–20, 2008.

[33] M. Jimichi. Exact moments of feasible generalized ridge regression estimator for linear basis function models. *Journal of Business Administration*, Vol. 60, No. 1, pp. 451–476, 2012. The Society of Business Administration, Kwansei Gakuin University (in Japanese).

[34] M. Jimichi. Exact moments of feasible generalized ridge regression estimator for linear basis function models. In *XXVIth International Biometric Conference*. International Biometric Society, August 2012.

[35] M. Jimichi and N. Inagaki. Centering and scaling in ridge regression. In K. Matsushita, M. Puri, and T. Hayakawa, editors, *Statistical Sciences and Data Analysis*, pp. 77–86. The Third Pacific Area Statistical Conference, VSP, 1993.

[36] M. Jimichi and N. Inagaki. r-k class estimation in regression model with concomitant variables. *Annals of the Institute of Statistical Mathematics*, Vol. 48, No. 1, pp. 89–95, 1996.

[37] I. T. Jolliffe. *Principal Component Analysis.* Springer-Verlag, second edition, 2002.

[38] M. G. Kendall. *A Course in Multivariate Analysis.* Griffin, London, 1957.

[39] S. Konishi and G. Kitagawa. *Information Criteria and Statistical Modeling.* Springer-Verlag, 2007.

[40] J. F. Lawless and P. Wang. A simulation study of ridge and other regression estimators. *Communication in Statistics. Theory and Method*, Vol. A5, pp. 307–323, 1976.

[41] T. S. Lee. Optimum ridge parameter selection. *Applied Statistics*, Vol. 36, pp. 112–118, 1987.

[42] T. S. Lee and D. B. Campbell. Selecting the optimum k in ridge regression. *Communications in Statistics. Theory and Method*, Vol. A14, pp. 1589–1604, 1985.

[43] D. V. Lindley and A. F. M. Smith. Bayes estimates for the linear model. *Journal of the Royal Statistical Society*, Vol. B34, pp. 1–41, 1972.

[44] C. L. Mallows. Some comments on C_P. *Technometrics*, Vol. 15, pp. 661–675, 1973.

[45] D. W. Marquardt. Generalized inverses, ridge regression biased linear estimation, and nonlinear estimation. *Technometrics*, Vol. 12, pp. 591–612, 1970.

[46] M. Nomura and T. Ohkubo. A note on combining ridge and principal component regression. *Communications in Statistics. Theory and Method*, Vol. A14, pp. 2489–2493, 1985.

[47] K. Ohtani. Distribution and density function of the feasible generalized ridge regression estimator. *Communications in Statistics. Theory and Method*, Vol. 22, pp. 2733–2746, 1993.

[48] C. R. Rao. *Linear Statistical Inference and Its Applications*. John Wiley and Sons, Inc., second edition, 1973.

[49] J. E. Rolph. Choosing shrinkage estimator for regression problems. *Communications in Statistics. Theory and Method*, Vol. A5, pp. 789–802, 1976.

[50] N. Sarkar. Comparison among some estimators in misspecified linear models with multicollinearity. *Annals of the Institute of Statistical Mathematics*, Vol. 41, pp. 717–724, 1989.

[51] G. A. F. Seber and A. Lee. *Linear Regression Analysis*. John Wiley and Sons, Inc., second edition, 2003.

[52] V. K. Srivastava and A. Chaturvedi. Some properties of the distribution of an operational ridge estimator. *Metrika*, Vol. 30, pp. 227–237, 1983.

[53] C. Stein. Inadmissibility of the usual estimator for the mean of a multivariate normal distribution. In *Proceedings of Third Berkeley Symposium on Mathematical Statistics and Probability*, Vol. I, pp. 197–206. University of California Press, Berkeley, 1956.

[54] E. M. Stein and R. Shakarchi. *Functional Analysis: Introduction to Further Topics in Analysis*. Princeton University Press, 2011.

[55] M. Sugihara and K. Murota. *Suuchi keisanhou no suuri*. Iwanami Shoten, 1994 (in Japanese).

[56] R. Tibshirani. Regression shrinkage and selection via the lasso. *Journal of the Royal Statistical Society. Series B (Methodological)*, Vol. 58, No. 1, pp. 267–288, 1996.

Index

C_L statistics, 39
r-k class, 7, 11
r-k-class-type, 14
g-inverse, 10

basis functions, 87
Bayesian methods, 39
beta function, 100
bias vector, 15
bootstrap, 73
Borel-Cantelli's lemma, 53

canonical form, 2
coefficient of determination, 6
conditional mean, 85

digamma function, 99

errors, 1
existence theorem, 35
explanatory variables, 1

feasible r-k class, 45
feasible GRR, 44, 85
feasible ORR, 43
feasible PCR, 45
FGRR, 85

gamma function, 99
Gauss-Markov setup, 2
Gaussian basis function, 88
generalized beta function, 100
generalized inverse, 10
generalized inverse regression, 11
generalized ridge regression, 7, 9
GRR, 9
GRR-type, 13

i.i.d., 1
independent and identically distributed, 1

information matrix, 2
intercept, 1
invariance, 4

LASSO, 86
LBF, 86
least absolute shrinkage and selection operator, 86
Lebesgue's dominated convergence theorem, 54
linear basis function, 86
linear basis function model, 87
linear basis function models, 86
linear basis function vector, 87

machine learning, 86
Markov's inequality, 52
MCE, 15
mean cross error, 15
mean squared error matirx, 15
mean structure, 86
method of good lattice points, 65
multicollinearity, 5, 86, 88

nearly linearly dependent, 5
normal equation, 3
normal linear model, 47, 74

OLS, 3
ordinary least squares, 3
ordinary ridge regression, 7, 9
ORR, 7, 9
ORR-type, 12
over-fitting, 86, 88

parametric bootstrap method, 74
parametric bootstrap OLS estimator, 75
parametric bootstrap shrinkage regression estimator, 76

INDEX

PCR, 7, 10
PCR-type, 13
polynomial regression model, 91
prediction sum of squares, 39
PRESS, 39
principal component regression, 7, 10
principal components, 3

radial basis function model, 88
regression coefficients, 1
regression function, 85
relative efficiency, 62, 70
response variables, 1
ridge coefficient, 9
ridge trace, 39

shrinkage estimators, 7
shrinkage regression estimators, 7, 9
standardized model, 2
statistical learning, 86
statistical regression problem, 85
stochastic part, 86
structural part, 86

TASE, 73
Taylor series expansion, 91
TMSE, 15
TMSEP, 39
total average squared error, 73
total mean squared error, 15
total mean squared error of prediction, 39
total variance, 5

usual linear regression model, 1

variance inflation factor, 6
variance-covariance matrix, 15
VIF, 6

weight vector, 87

著者略歴

地道 正行（じみち まさゆき）

兵庫県出身
神戸商科大学商経学部管理科学科卒業
大阪大学大学院基礎工学研究科数理系専攻修士課程修了
現在, 関西学院大学商学部教授

関西学院大学研究叢書　第 182 編

Shrinkage Regression Estimators and Their Feasibilities

2016 年 11 月 25 日初版第一刷発行

著　者　　地道正行

発行者　　田中きく代
発行所　　関西学院大学出版会
所在地　　〒 662-0891
　　　　　兵庫県西宮市上ケ原一番町 1-155
電　話　　0798-53-7002

印　刷　　株式会社遊文舎

©2016 Masayuki Jimichi
Printed in Japan by Kwansei Gakuin University Press
ISBN 978-4-86283-228-3
乱丁・落丁本はお取り替えいたします。
本書の全部または一部を無断で複写・複製することを禁じます。